APPLICATIONS
OF POLYMERS

APPLICATIONS OF POLYMERS

Edited by

Raymond B. Seymour

University of Southern Mississippi
Hattiesburg, Mississippi

and

Herman F. Mark

Polytechnic University
Brooklyn, New York

PLENUM PRESS • NEW YORK AND LONDON

Library of Congress Cataloging in Publication Data

Applications of polymers.

"Proceedings of the American Chemical Society symposium honoring O. A.
Battista, on Applied Polymer Science, sponsored by Phillips Petroleum Com-
pany, held April 9, 1987, in Denver, Colorado"—T.p. verso.
    Includes bibliographies and index.
    1. Polymers and polymerization—Congresses. 2. Battista, O. A. (Orlando
Aloysius), 1917-   —Congresses. I. Seymour, Raymond Benedict, 1912-   .
II. Mark, H. F. (Herman Francis), 1895-   . III. Battista, O. A. (Orlando
Aloysius), 1917-   . IV. American Chemical Society. V. Phillips Petroleum
Company.
QD380.A65   1987   547.7   87-29154
ISBN-13: 978-1-4684-5450-5   e-ISBN-13: 978-1-4684-5448-2
DOI: 10.1007/978-1-4684-5448-2

Proceedings of the American Chemical Society Symposium honoring
O. A. Battista, on Applied Polymer Science, sponsored by Phillips Petroleum
Company, held April 9, 1987, in Denver, Colorado

FOREWORD

Natural polymers, such as proteins, starch, cellulose, hevea rubber, and gum which have been available for centuries, have been applied as materials for food, leather, sizings, fibers, structures, waterproofing, and coatings. During the past century, the use of both natural and synthetic polymers has been expanded to include more intricate applications, such as membranes, foams, medicinals, conductors, insulators, fibers, films, packaging and applications requiring high modulus at elevated temperatures.

The topics in this symposium which are summarized in this book are illustrative of some of the myriad applications of these ubiquitous materials. As stated in forecast in the last chapter in this book, it is certain that revolutionary applications of polymers will occur during the next decades. Hopefully, information presented in other chapters in this book will catalyze some of these anticipated applications.

It is appropriate that these reports were presented at an American Chemical Society Polymer Science and Engineering Division Award Symposium honoring Dr. O.A. Battista who has gratifying to note that Phillips Petroleum Company, which has paved the way in applications of many new polymers, is the sponsor of this important award.

We are all cheerfully expressing our thanks to this corporate sponsor and to Distinguished Professor Raymond B. Seymour of the University of Southern Mississippi who served as the organizer of this symposium and editor of this important book.

H. Mark

CONTENTS

COMMERCIALIZATION OF NEW COLLOIDAL POLYMER MICROCRYSTALS*

O. A. Battista

The O. A. Battista Research Institute
3863 SW Loop 820, Suite 100
Fort Worth, Texas 76133

Commercial uses of single colloidal polymer microcrystals were first demonstrated in 1955, using cellulose as the raw materials. A worldwide, multi-million, pounds-per-year industry emerged from this discovery. The market is still growing in annual demand more than 30 years later. Today, this specialized field of polymer science is undergoing a major new surge of commercialization based on past as well as never-before-available colloidal polymer microcrystals.

Expanded commercialization of single colloidal-sized polymer microcrystals is underway. Production has already progressed past the pilot-plant stages for polymer microcrystals other than cellulose microcrystals.

One particularly new broad-based opportunity that is being developed has to do with the conversion of hundreds of millions of pounds of waste polymer products into new colloidal aqueous suspensoids or dry colloidal particles. Some of these new forms have greater value than their natural or man-made polymer precursors.

For example, conservative estimates based on 1986 data project that at least 700,000,000 pounds of waste polymer products will be available in the U. S. A. alone as a reservoir of acceptable, reusable, polymer, raw materials. The toxicity and environmental hazards created by the incineration of waste polymer materials have become of national concern. The cost of burying them in "graveyards" many miles from the site of their origin has been estimated conservatively at 15 cents a pound! A new industry designed to give a "second life" to these valuable natural materials and man-made waste polymer forms – an industry to recycle particular waste polymer forms – is about to leave the heel of an exponential growth curve.

---

*American Chemical Society's Award Lecture, Applied Polymer Science Award sponsored by Phillips Petroleum Co., delivered on April 6, 1987, at the 193rd National Meeting in Denver, Colorado.

What are the raw materials of primary commercial interest?

They are: waste cellulose (rayon, cellophane, alpha cellulose, and cotton), nylons (especially nylon -6 and nylon -66), polyesters (in fibrous, textile, and structural forms), crystalline amylose, and even silk and wool wastes! In addition, millions of pounds of valuable discarded cotton, nylon, and polyester fabrics are hanging in the household closets of America - not to mention all of the rest of the world.

From the late thirties through the discoveries of Carothers and countless others up until today, conventional polymer science has focused its thrust largely on making molecules longer and from which stronger and tougher fibers, films, or plastic structures could be manufactured. <u>Conventional polymer science has been remarkably successful</u>. It has spawned products with sales in excess of $70,000,000,000 a year in the United States alone.

Microcrystal polymer science is the antithesis of more conventional polymer chemistry. It requires using as raw materials fibers, films, resins, or plastics that have already had the "genetics" of their microcrystalline regions fixed in place with predetermined dimensions and degrees of perfection. Carefully controlled, selective chemistry is used to break up and/or loosen the more disordered networks of interconnecting molecules within which polymer microcrystals are embedded in a molecular matrix fixed by linear, covalent, and molecular bonds. After using chemistry to disconnect the covalent, molecular linkages between regions of high molecular order (microcrystallites) and interconnecting regions of lesser molecular order, appropriate mechanical energy is used to provide high sheer forces to liberate the polymer microcrystals into single virus-sized, highly-crystalline particles. This process requires that the chemical treatment must be controlled carefully so that the size of the microcrystals will remain as close as possible to their dimensions as they existed in the precursor polymer raw materials. The "liberated" polymer microcrystals, depending on their respective precursor raw material, may range in size from 4,000 angstroms in maximum dimensions to as small as 200-300 angstroms.

The commercial potential of recycling waste polymer materials in colloidal forms is not a "pie-in-the-sky" projection. Producing countless millions of pounds of microcrystalline cellulose from a single polymer raw material - wood pulp alpha cellulose - is a proven reality worldwide. You would be hard put to find anyone in the food and/or pharmaceutical industry from New York to Paris to Moscow or to Tokyo who has not used or at least heard of AVICEL Microcrystalline Cellulose. A recent computer printout of literature references about microcrystalline cellulose stunned me! It contained about 4,000 separate publications. More than ten new plants have been built in recent years to manufacture cellulose microcrystals alone.

Even so, as a consultant to eight major polymer-oriented corporations, I am constantly puzzled by the almost naive understanding that many chemists still have of the science of colloidal polymer micro-crystals.

2

Two unexplored avenues for each of the newer colloidal polymer microcrystal species are:

1) Commercialization of much <u>smaller</u> microcrystals by selective use of new and/or chemically modified precursor raw materials and

2) Topochemical derivatization at very low degrees of substitution of derivatives of polymer microcrystals.

The emerging new technology will provide increasing market opportunities for colloidal water suspensoids and/or dry, colloidal aggregates or clusters of submicron, single, near-perfect, polymer microcrystals.

Among the industrial end uses being evaluated are non-toxic, water-based systems of binders and coatings for paper; board and pulp molded products, including particle and chip boards for the building industries; prime coatings for glass, aluminum, wood, ferrous, and other non-ferrous surfaces; thickeners for water-based paints; etc. In dry form, the colloidal microcrystals may be used for fluid bed and electrostatic coatings, as binders for industrial and agricultural tableting, and as additives for changing the characteristics of existing molding and casting polymers. Some of the more unusual uses will be in concrete products, synthetic ivory, and catalytic substrates - to name a few.

The first commercial plant for recycling waste polymers as single polymer microcrystals is expected to be on-stream in the latter months of 1987. Projections are that the recycling of waste cellulose, nylon, polyester fibers and/or plastics, etc. will spawn a new worldwide industry that will introduce second life products at a profit and, at the same time, reduce the environmental hazards of present ways of disposing of them. This plant will be constructed and operated by MCP's, INC., a subsidiary of MICROTECH INDUSTRIES, INC., Suite 201, 275 MacPherson Avenue, Toronto, Ontario, Canada M4V 1A4.

On the 32nd anniversary of the discovery of commercial uses for polymer microcrystals, the emerging new industry we have described proffers to make a significant future impact in broadening the opportunities of polymer chemistry and in making important improvements to the quality of life on a global scale. It has been an honor and a privilege to have spent over 40 years of my professional career in this science. I am most grateful to each participant in this Symposium and to the Phillips Petroleum Company, sponsor of the Applied Polymer Science Award.

FIBERS

Charles H. Fisher

Chemistry Department
Roanoke College
Salem, Virginia 24153

Fibers,[1-7] flexible and having a high ratio of length to width and cross section, are an important segment of the 100 million tons of polymers manufactured annually in the world.[8] Fibers have been classified[3] as:

Naturally Occurring Fibers

> Vegetable (cellulosic), e.g., cotton, linen, ramie.
> Animal (protein), e.g., wool, mohair, silk.
> Mineral, e.g., asbestos.

Man-Made or Manufactured Fibers

> From natural organic polymers, e.g., rayon, acetate.
> From synthetic organic polymers, e.g., polyester, nylon.
> From inorganic substances, e.g., glass, metallic, ceramic.

The production of natural fibers continues to be a major agricultural activity world-wide; the production of man-made or chemical fibers is a major activity in the chemical industry. The textile and paper industries are the primary converters of fibers into the numerous products needed by five billion humans in our modern society.

Fibers and fibrous products, like food, were critically important to the some 113 billion humans who lived and died over the past 2.5 million years.[9] The fibrous materials of prehistory and ancient times included hides, furs, cords, ropes, mats, nets and baskets. Tying knots might have been an early invention. Inventions of much greater importance, spinning and weaving, might have occurred about 35 thousand years ago; this could be inferred from patterns of weaves found on clay vessels of the Old Stone Age.[6]

The earliest evidence of woolen textiles dates from about 6000 B.C. Bits of linen from Egypt indicate that people there wove flax about 5000 B.C. By 3000 B.C., cotton was grown in the Indus River Valley in what are now Pakistan and western India. The Chinese began to cultivate silkworms about 2700 B.C.[1,6,10]

Advances in technology have had a dramatic impact on the production and processing of fibers and fibrous products. Man-made or chemical fibers, manufactured from synthetic polymers, have replaced the natural fibers in many uses; they also are the basis for new products and new uses.

Table 1. Fiber Consumption by the U.S. Textile Industry, 10^6t[a]

Fiber Type	1976	1980	1984
cotton	1.54	1.41	1.21
wool	0.04	0.04	0.07
rayon	0.27	0.23	0.18
acetate	0.14	0.14	0.09
polyester	1.50	1.59	1.45
nylon	0.95	1.04	1.09
acrylic	0.27	0.27	0.23
polyolefin	0.23	0.32	0.45
glass	0.04	0.09	0.09
Total	4.98	5.13	4.90

[a]Refs. 3 and 7.

The production of manufactured fibers has greatly increased in both the U.S. (Table 1) and the world (Table 2).

Textile products now include woven and knitted goods, draperies, blankets, towels, felts, laces, nets, braids, and an incredible variety of fabrics. In the United States, the textile industry manufactures about

Table 2. World Fiber Production 10^6t[a]

Fiber type	1974	1976	1978	1980	1982	1984
Natural fiber						
cotton	14.05	12.48	12.97	13.99	14.63	16.48
wool	1.53	1.49	1.53	1.61	1.63	1.67
silk	0.04	0.05	0.05	0.06	0.06	0.06
Synthetic fiber						
cellulosics	3.53	3.21	3.32	3.24	2.94	3.08
synthetic polymer	8.20	9.45	11.04	11.56	11.35	13.19

[a]Refs. 3 and 7.

25 billion square yards (21 billion square meters) of fabric a year. About 70 percent of this output is used in making clothing and household goods.[1] Textiles are also used in thousands of other products. These products include basketball nets, boat sails, bookbindings, conveyor belts, fire hoses, flags, insulation materials, mailbags, parachutes, typewriter ribbons, and umbrellas. Automobile manufacturers use fabrics in the carpeting, upholstery, tires, and brake linings of cars. Hospitals use adhesive tape, bandages, and surgical thread. Surgeons replace diseased heart arteries with arteries knitted or woven from textile fibers.[1]

Natural Fibers

Only natural fibers were available during most of history. Natural fibers--from plants, animals, and minerals--still account for more than half the fibers produced annually in the world.

Cotton, the most widely used natural fiber, is processed into numerous apparel, home furnishings, and industrial products. Flax, a strong, silky fiber from the stems of flax plants, is used in making clothing, napkins and other linen products. Hemp, jute, and sisal are coarse fibers used in cords, ropes, and rough fabrics.

Animal fibers include fur and hair. Wool, the hair sheared from sheep and certain other animals, is popular in clothing and home furnishings. Rough surfaces on wool fibers give bulk and warmth to wool clothing and blankets. Silk is the strongest natural fiber. Manufacturers unwind silk filaments from silkworm cocoons and make silk yarn for clothing and household fabrics.

Various minerals--chiefly serpentine and amphibole--that occur in fibrous form are called asbestos. They resist high temperatures and are used in insulation, shingles, and fireproof products.

Man-made or Manufactured Fibers

The chemical or manufactured fibers (Tables 3 and 4), products of 20th century technology, account for more than two-thirds of the fibers processed today in U.S. textile mills. Manufactured fibers consist of two broad groups. The first group, of which rayon and acetate are examples, are produced by modifying natural polymers such as cellulose. The second group, frequently called synthetic or man-made, includes fibers such as polyester and nylon that are made from chemical intermediates. Unlike most natural fibers, manufactured fibers are produced in long, continuous lengths called filaments. Many manufactured fibers also have certain qualities superior to those of natural fiber. The most widely-used manufactured fibers are the polyesters, nylons, polyacrylics, and polyolefins.

Rayon fibers, particularly staple fibers, serve in a wide variety of uses, alone and in combination with other fibers. High-tenacity, high total tex (denier) filament is used in tire cord and other industrial uses. Rayon staple is produced in a variety of types, including polynosic and high wet-modulus types, these retaining greater strength and dimensional stability when wet. Rayon staple is used in apparel, household goods, and various nonwoven fabrics.

Cellulose acetate fabrics have attractive appearance, pleasant hand, and excellent thickness. Women's apparel, draperies, and upholstery are major uses. Cigarette filters represent a major and still growing use for acetate tow.

Table 3. Generic Names for Manufactured Textile
Fibers[a],[b]

Generic name	Definition of fiber-forming substance[a]
acetate	cellulose acetate; triacetate where not less than 92% of the cellulose is acetylated
acrylic	at least 85% acrylonitrile units
aramid	polyamide in which at least 85% of the amide linkages are directly attached to two aromatic rings
azlon	regenerated naturally occurring proteins
glass	glass
modacrylic	less than 85% but at least 35% acrylonitrile units
novoloid	at least 85% cross-linked novolac
nylon	polyamide in which less than 85% of the amide linkages are directly attached to two aromatic rings
nytril	at least 85% long chain polymer of vinylidene dinitrile where the latter represents not less than every other unit in the chain
olefin	at least 85% ethylene, propylene, or other olefin units
polyester	at least 85% ester of a substituted aromatic carboxylic acid, including but not restricted to substituted terephthalate units and para-substituted hydroxybenzoate units
rayon	regenerated cellulose with less than 15% chemically combined substituents
saran	at least 80% vinylidene chloride
spandex	elastomer of at least 85% of a segmented polyurethane
vinal	at least 50% vinyl alcohol units and at least 85% total vinyl alcohol and acetal units
vinyon	at least 85% vinyl chloride units

[a]All percentages are by weight.

[b]Ref. 4.

Table 4. Man-made Fibers: Their Properties And Uses[a]

Fiber	Trade Names	Characteristics	Uses
Acetate	Acele, Celaperm, Estron	Resists mildew, shrinking, stains, and stretching	Clothing; draperies; upholstery
Acrylic	Acrilan, Creslan, Orlon, Zefran	Soft; resists mildew, sunlight, and wrinkling	Blankets; carpeting; clothing; upholstery
Aramid	Kevlar, Nomex	Resists heat, chemicals, and stretching	Bulletproof vests; electrical insulation; rope; tires
Glass	Beta, Fiberglas, PPG, Vitron	Resists chemicals, flames, mildew, moisture, and sunlight	Draperies; electrical insulation; ironing board covers
Metallic	Brunsmet, Chromeflex, Fairtex, Lurex, Metlon	Resists insects, mildew, and tarnishing	Decorative trim for bedspreads, table-cloths, and upholstery
Moda-crylic	Verel	Soft; resists chem-icals, flames, and wrinkling	Artificial furs; blankets; carpeting; wigs
Nylon	Antron, Cumuloft, Enkaloft, Quiana	Strong; elastic; easy to launder; dries quickly; retains shape	Carpeting; hosiery; lingerie; parachutes; upholstery
Olefin	DLP, Herculon, Marvess, Vectra	Light-weight; resists insects, mildew, moisture and sunlight	Automobile seat covers; filters; indoor-outdoor car-peting
Polyester	Dacron, Encron, Fortrel, Kodel	Resists wrinkling; easy to launder; dries quickly	Blankets; carpeting; clothing; fire hose; sewing thread
Rayon	Avril, Fibro	Absorbent; easy to launder; dyes easily	Carpeting; clothing; draperies; upholstery
Rubber (Synthetic)	Contro, Lactron, Lastex	Strong; elastic; repels moisture	Mattresses; support hose; swimwear; underwear
Saran	Rovana, Velon	Resists acids, insects, mildew, moisture, and stains	Draperies; outdoor furniture; rainwear; upholstery
Spandex	Glospan, Lycra, Numa	Elastic; lightweight; resists sunlight and and perspiration	Fitted sheets; slip-covers; support hose; swimwear; underwear
Tri-acetate	Arnel	Resists shrinking, stains, and wrinkling; dries quickly	Draperies; sportswear; blended with other fibers

[a]Ref. 1.

Cellulose triacetate filament became a useful specialty fiber in the 1950s. Unlike regular cellulose, it can be heat-treated to induce a degree of crystallization, which produces dimensional stability and related "ease-of-care" characteristics. As a result heat-treated triacetate resembles such fully synthetic fibers as nylon or polyester.

Alginate and protein fibers are two additional textile products that have been made from natural polymers. Their commercial success has been limited.

Polyester fibers are now the largest-volume man-made or chemical fiber. These are manufactured from poly(ethylene terephthalate) and from the polymer made by condensing terephthalic acid with 1,4-dimethylolcyclo-hexane. The latter type of fiber melts higher than the first-mentioned polyester and has a lower specific gravity and excellent recovery from stretch.

Polyester fibers are produced in both staple and filament form. The fiber is remarkably versatile. It is strong, abrasion resistant, relatively stable, higher in modulus than nylon, and of lower moisture regain.

Uses range widely over the apparel, home furnishings, automotive, and industrial fields. Enormous quantities of polyester staple are blended with cotton, rayon, or wool in spun yarns used for apparel. Good abrasion resistance also suits polyester to use in carpeting. Continuous filament yarn finds wide use in apparel; much of it is textured to produce bulk and opacity for this use.

Polyester is used in tire cord. Other industrial uses include belts, ropes, and filter fabrics.

The principal aliphatic polyamides are nylon-6,6 (made from adipic acid and hexamethylenediamine) and nylon-6 (made from caprolactam).

Nylon-6,6 is a strong, tough, abrasion-resistant fiber that is relatively stable and relatively readily dyeable. Nylon-6,6 has a wide spectrum of uses. Textile nylon is used in hosiery, apparel, and home furnishings. Most of this fiber is multifilament, but monofilament nylon is used in sheet women's stockings. The high strength of nylon permits manufacture of lightweight and very sheer fabrics. Because of good abrasion resistance and resilience, nylon dominates the U.S. carpet market. Strong, durable nylon monofilaments are used in fishing lines and fishnets; high-strength nylon multifilament has a variety of automotive and industrial uses ranging from belting to filter fabrics. Tire cord is a major use.

The fiber properties and uses of nylon-6 are generally similar to those of nylon-6,6.

Quiana, a polyamide containing alicyclic rings, is silk-like in appearance and resilience. Quiana fabrics are used mainly in apparel, particularly dresses, blouses, and shirts.

Aromatic polyamides (aramid fibers) are made from aromatic inter-mediates. They are high melting, high in thermal stability, and generally possessed of high performance properties. The aramid fibers are higher in price, and find use in applications where exceptionally high strength, high modulus or high resistance to heat, or both, are required.

Commercial aramid fibers are trademarked Nomex and Kevlar by DuPont. Nomex, based on meta-linked isophthalic acid and m-phenylenediamine, is used in specialty papers for electrical insulation and aircraft structures, in protective garments, and in other applications requiring high thermal stability. Kevlar is an aromatic polyamide containing para linkages. It melts at over 500°C, is expectionally high in strength with a tenacity more than twice that of high strength nylon or polyester, and a very high modulus. It finds use in heavy duty conveyor belts, and in composite structures with casting resins such as epoxies. It competes with steel in radial tire reinforcement.

Acrylic fibers, based primarily on acrylonitrile, have good tenacity, although less than polyester or polyamides, excellent stability to sunlight, good dye acceptance as a result of the copolymer system used for this purpose, and a soft, pleasing hand of wool-like characteristics. Abrasion resistance, although well below that of nylon or polyester, is nevertheless good, and superior to that of wool. The acrylics find widespread use in both indoor and outdoor furnishings, including awnings and draperies, and in blankets, sweaters, and carpets. Blending low- and high-shrinkage yarns and shrinking the blend produces a bulky yarn.

The modacrylic fibers are acrylonitrile copolymers or terpolymers with significant comonomer contents. Among the comonomers used are vinylidene chloride and vinyl chloride. The fiber-forming modacrylic polymers are more soluble than the acrylics and have various textile uses, including fake fur pile fabrics and protective garments. Modacrylics, because of their flame-resistant characteristics, are used also in wigs and doll's hair.

Polypropylene fiber is relatively low melting for a chemical fiber. Offsetting this and other relative deficiencies are low cost, high strength, great chemical inertness, and, because of the low density, high yardage of fiber of a given tex per kilogram (den/lb). Because of the chemical inertness, dye acceptance is inadequate. For uses where color is requisite, either the base polymer is modified to provide dyeability or the fiber is spun-colored. Because of oxygen and light sensitivity, suitable stabilizers are incorporated. Textile uses include upholstery and carpeting, particularly indoor-outdoor carpets. In addition to face yarns for carpeting, polypropylene fiber, particularly that produced from film, is used for woven carpet backings. The low melting point and the fabric hand have largely precluded polyolefins from application in apparel fabrics. Other uses include rope and cordage, fishnets, and filter media. Negligible moisture adsorption, resistance to decay by organisms, and low density, which causes the fiber to float make polyolefins particularly suited for some uses.

Man-made elastomeric fibers, e.g., Spandex, differ from the usual textile fibers in having high extensibility to break (500-600%) and high recovery from stretching. The fibers, white and dyeable, are stronger and lighter than rubber; they are particularly suitable for use in foundation garments, bathing suits, support hose, and other elastic products.

Vinyon fibers are about 85-90% vinyl chloride and 10-15% vinyl acetate units. The fiber is temperature sensitive, starting to soften, shrink, and become tacky below the boiling point of water. It finds specialty use in applications requiring bonding and heat sealing, in conjunction with other fibers.

Vinyon fibers from vinyl chloride homopolymers have relatively limited thermal stability, tending to shrink at temperatures in the neighborhood of boiling water. They are, however, resistant to moisture and rotting, and inherently nonflammable, suiting them to specialty uses such as filter cloths, nonflammable garments, fishnetting, and felts for insulating purposes.

Saran fibers are based on vinylidene chloride copolymerized with small amounts of vinyl chloride, and still smaller amounts of acrylonitrile. They are characterized by a pale straw color and by resistance to water, fire, and light, and bacterial and insect attack. Their relatively low melting points require excessively low ironing temperature. The fibers find specialty use in certain types of upholstery, filter cloths, and fishnets. The relatively low cost of these fibers is one of their attractions.

PeCe fiber is based on a post-chlorinated vinyl chloride polymer. The post-chlorination raises the chlorine content of the polymer from 57 to 64% and confers acetone solubility. Like many other vinyl fibers, it is low melting and restricted to special applications.

Vinal fibers, based on poly(vinyl alcohol) have reasonable strength, a moderately low melting point (222°C), limited elastic recovery, good chemical resistance, and resistance to degradation by organisms. Vinal is used in bristles, filter cloths, sewing thread, fishnets, and apparel. Production has remained largely in Japan.

Polychal fibers are related to the Vinyon and Vinal types. Polychal fibers have been exported from Japan to the U.S. for use in flame-retardant apparel.

Teflon fibers, based on polytetrafluoroethylene, have uniquely high chemical stability and inertness, no water absorption, low frictional characteristics, and high melting points with decomposition preceding and accompanying melting. Teflon fibers in both filament and staple forms are high priced and find, as would be expected, highly specialized applications, such as packings, special filtration fabrics, and other uses where corrosion resistance, lubricity, and temperature resistance are required.

Kynol, a phenolic-type fiber having fire-resistant properties, is used in protective garments.

Polybenzimidazole fibers are of the high performance type, of very high melting point, good strength and extensibility, nonflammable, and surprisingly for a synthetic fiber, of high moisture regain: 13%. These fibers are of interest for aerospace and industrial applications.

Poly(phenylene sulfide) fibers, produced by the reaction of p-dichlorobenzene with sodium sulfide, are used as engineering thermoplastic polymer. These fibers have good dimensional stability, flame resistance, thermal stability, and chemical resistance. They are used in air filtration applications and as conveyor belts in high temperature drying operations.

Benzoate fibers are made in Japan by the self-condensation of p-(beta-hydroxyethoxy) benzoic acid to provide silk-like products.

Poly(hydroxyacetic acid) or poly(glycolic acid) fibers are crystalline, and can be oriented by stretching. Because they are absorbable in the body, they have become important in surgical suture applications, where they replace catgut sutures in many such applications.

Polyphosphazene fibers are being developed; the anticipated applications include biomaterials, drug delivery systems, electronics, and separation membranes.[11]

Glass fibers are inorganic, strong, nonflammable, and rather heat-resistant, as well as highly resistant to chemicals, moisture, and attack by organisms. They are low in extensibility and higher in density than the organic fibers. To offset their brittleness, glass fibers are spun into fine filaments. After lubrication, these are wound as a multifilament strand unless insulating batting is being produced; in the latter case the filaments are attenuated while still molten and are deposited in the form of a thick bat.

Glass fibers, in forms ranging from filament yarn to mats, woven rovings, and staple, find a variety of important uses. In addition to use as insulating bats, they are used in fabric form for draperies where the nonflammability, inertness, and resistance to the effects of sunlight are assets. Glass fiber cloth, as well as batting, is used in a variety of insulating applications and for filtration.

One of the important uses for glass fiber is in reinforced plastics, particularly reinforced thermosetting polyester resins. Such composites are widely used in industrial and automotive applications. Glass-filled thermoplastic resins have also grown in volume. The glass fibers are used in nylon, polyacetal, and poly(butylene terephthalate) molding resins. Glass fibers are used in both radial and bias-ply automotive tire reinforcement.

Metallic and various inorganic fibers have been studied for special applications. Fine steel wire is used for radial tire reinforcement. Metallic fibers have been used in small quantities with organic fiber carpet yarns to minimize static. Fibers of boron, boron-tungsten, steel, beryllium, boron and silicon carbides, boron and silicon nitrides, alumina, zirconia, and other inorganics have received attention for high performance specialty uses. Most are costly, and the quantities employed are small.

Carbon fibers are characterized by extremely high strength and modulus along with high temperature resistance. These fibers are used in high performance, reinforced composite structures where strength, stiffness, and lightness of weight are at a premium, e.g., in special aerospace, industrial, and recreational applications. The composite structure may include other fibers, such as glass, along with a matrix resin, such as polyesters, epoxies, or polyimides.[12]

Composite materials containing fibers are used in applications in which high strength and low weight are required, as in aeronautics. Composites are heterogeneous structures consisting of a continuing matrix in which fibers are dispersed and embedded. The reinforcing fibers add tensile strength and dimensional stability. Some of the products made from composites are airplanes, boats, auto parts, chemical process equipment, appliance cabinets, machine housings, and caskets.[13]

REFERENCES

1. D. S. Hamby, Vol. 19, pp. 168-173 in World Book Encyclopedia, World Book, Inc., Chicago (1987).
2. W. J. Roberts, pp. 471-472 in Kirk-Othmer Concise Encyclopedia of Chemical Technology, John Wiley & Sons, Inc., New York (1985).
3. Ludwig Rebenfeld, Vol. 6, pp. 647-733 in Encyclopedia of Polymer Science and Engineering, 2nd Ed., John Wiley & Sons, Inc. (1986).
4. W. J. Roberts, Vol. 10, pp. 148-166 in Kirk-Othmer Encyclopedia of Chemical Technology, 3rd Ed., John Wiley & Sons, Inc., New York.
5. J. C. Arthur, Jr., Vol. 4, pp. 261-284 in Encyclopedia of Polymer Science & Engineering, John Wiley & Sons, Inc., New York (1986).
6. R. B. Seymour, Editor, History of Polymer Science and Technology, Marcel Dekker, Inc., New York (1982).

7.   Textile Organon, <u>56</u>, (1985).
8.   G. E. Zaikov, Chemistry International, (9), 89 (1987).
9.   Thomas Exter, Atlantic Magazine, Apr. 1987, p. 8.
10. M. L. Ryder, Scientific American, Jan. 1987, pp. 112-119.
11. Chemical Week, Mar. 25, 1987, p. 5.
12. Chemical Week, Apr. 29, 1987, p. 17.
13. W. Worthy, Chem. & Eng. News, Mar. 16, 1987, pp. 7-13.

HIGH MODULUS POLYMERS

Raymond B. Seymour

Department of Polymer Science
University of Southern Mississippi
Hattiesburg, Mississippi 39406-0076

Professor R. S. Porter of the University of the Massachusetts, who is also an adjunct professor in the Department of Polymer Science at the University of Southern Mississippi, presented a report on the generation and analysis of Polymer Morphologies and High Chain Extension at the Phillips Symposium. Tensile properties of semicrystalline thermoplastics may be improved by control of crystallization temperature and by the ultradrawing of polymer to produce polyolefins with tensile moduli of 220 GPA or polyethylene and 33 GPa for polypropylene.

The voids which may form as a result of stretching of these polymers, may be eliminated by the use of a compression rolling process using 2 rolls rotating in opposite directions at equal velocity. The need for expensive rolling equipment has been reduced by adding a liquid to decrease the friction between the roll and the polymer sheet. Oriented high density polyethylene sheets with a secant modulus of 12 GPa have also been obtained by use of a differential speed compression rolling process combined with high temperature stretching.

ASYMMETRIC POLYSULFONE HOLLOW FIBER

MEMBRANES FOR GAS SEPARATIONS

A.K. Fritzsche

Permea, Inc., A Monsanto Company
11444 Lackland Rd.
St. Louis, MO 63146

INTRODUCTION

Loeb and Sourirajan invented the first integrally-skinned membrane in 1960 for desalination by phase inversion of cellulose acetate sols (1). In the integrally-skinned membrane, the skin and substructure are composed of the same material. The skin layer determines both the permeability and selectivity of the bilayer, whereas the porous substructure functions primarily as a physical support for the skin. Differences in structure between the two layers are the result of interfacial forces and the fact that solvent loss occurs more rapidly from the air-solution and solution-nonsolvent interfaces than from the solution interior (2).

Integrally-skinned hollow fiber membranes suitable for commercial gas separations were subsequently developed by Monsanto by application of a multicomponent membrane (3). For example, a thin layer of a highly permeable, nonselective polymer was applied to the surface of an asymmetric polysulfone hollow fiber membrane. This coating sufficiently reduces the permeability through the pores and defects within the skin to render permeation through the normal skin layer predominant (4).

It is the object of this paper to review the morphology of asymmetric hollow fiber membranes and the influences of this morphology on membrane transport rate and selectivity.

CALCULATION OF PERMEABILITY AND SELECTIVITY

The permeability is the rate of transport of a gas through a membrane per unit time. It is expressed as

$$P = \frac{cm^3(STD) - cm}{cm^2 - sec - cmHg} \qquad (1)$$

In an asymmetric membrane the thickness (ℓ) of the effective separating layer is unknown. So, flux measurements are given in terms of the P/ℓ and determined by equation 2.

$$P/\ell = \frac{Q}{A\Delta P} = \frac{Q}{n\pi DL\Delta P} \qquad (2)$$

where

P = the permeability of the separating layer $(cm^3(STD)-cm)/$
  $(cm^2-sec-cmHg)$,

ℓ = the effective thickness of the separating layer,

Q = the gas flux,

n = number of fibers in the sampling,

D = outer diameter of the hollow fiber membrane,

L = active length of the fibers, and

ΔP = differential pressure between the outer surface and
  bore of the hollow fiber membrane.

The selectivity, also referred to as the separation factor, of a gas pair is the ratio of the permeation rates. For a gas pair, such as helium and nitrogen, the separation factor is given by equation 3.

$$\alpha = \frac{P/\ell \ (He)}{P/\ell \ (N_2)} \tag{3}$$

It is evident from equations 2 and 3 that the permeability of a membrane depends on the thickness of the effective separating layer while the selectivity is independent of the separating layer thickness.

MORPHOLOGY OF ASYMMETRIC HOLLOW FIBER MEMBRANES

The advantage of an asymmetric membrane is that the effective separating layer is only a small fraction of the total membrane wall thickness. The remaining structure in the wall serves only to support this thin outer layer. Polysulfone hollow fiber membranes can be fabricated by dissolving polysulfone in solvents such as dimethylformamide, dimethylacetamide, N-methylpyrrolidone and formylpiperidine and by spinning into a coagulation medium, such as water (2,3). The solvent is miscible in the coagulation medium, and the polymer is not. Often, small amounts of nonsolvent, such as formamide, are added to the spinning dope to alter the kinetics of coagulation and the resultant membrane structure (2,3). Polysulfone hollow fiber membranes prepared from the formyl-piperidine/formamide solvent/nonsolvent mixture are typical of this art and are used as examples in this discussion. The structure is illustrated in Figure 1, which is a SEM photomicrograph of a cross-section of an asymmetric

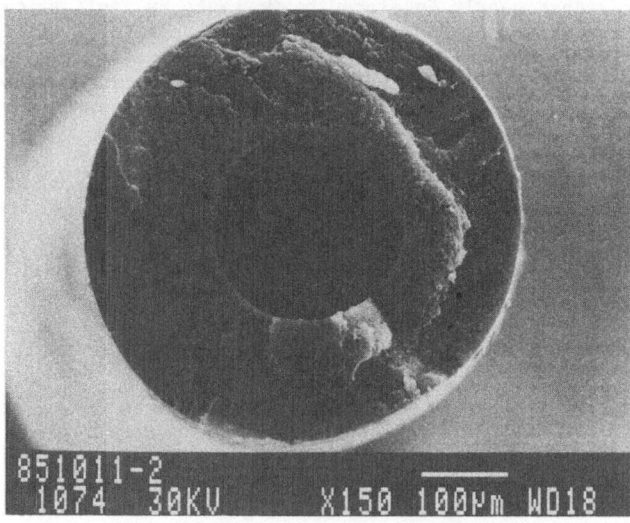

Figure 1.   SEM of Cross-section of Asymmetric Polysulfone Hollow Fiber Membrane Spun from Formylpiperidine/Formamide.

polysulfone hollow fiber membrane spun from this solvent/nonsolvent mixture. This SEM photomicrograph shows a polysulfone hollow fiber membrane with a wall thickness of approximately 135 μm. At this resolution the porous internal structure of the hollow fiber membrane can be barely discerned while the outer skin containing the effective separating layer cannot be detected. The outer edge of this fiber membrane at higher magnification is shown in Figure 2.

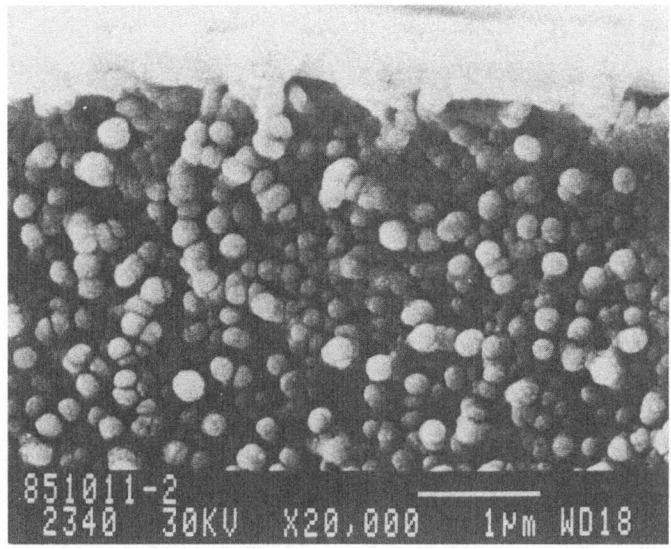

Figure 2. SEM of Outer Edge of Asymmetric Polysulfone Hollow Fiber Membrane Spun from Formylpiperidine/Formamide.

Figure 3. SEM of Interior Matrix of Asymmetric Polysulfone Hollow Fiber Membrane Spun from Formylpiperidine/Formamide.

19

TABLE I

Performance of Uncoated Polysulfone Hollow Fiber
Membranes Spun from Formylpiperidine/Formamide as
Function Oxygen Plasma Exposure Time

Etch Time (min)	$P/\ell(He)^{a}$ $\times 10^6$	$P/P(N_2)^{a}$ $\times 10^6$	α
0	62.9	4.7	13.4
.5	47.9	4.1	11.9
1.0	57.1	5.8	9.6
1.5	95.9	21.9	4.4
2.0	153	43.3	3.4
3.0	173	59.1	3.0
5.0	203	71.4	2.9
10.0	174	124	1.4

a. All permeability units are $cm^3(STD)/cm^2$-sec-cmHg

A dense skin approximately .5-.7 μm (500-700nm) exists at the outer surface of the hollow fiber membrane. This skin is composed of micelles so tightly packed that it is difficult to identify their boundaries. This skin contains the effective separating layer of the hollow fiber membrane and micropores with sizes which are below the limits of resolution of the scanning electron microscope. Below the skin, a region of tightly packed spherical micelles with readily discernible boundaries can be observed. These micelles become less tightly packed as the distance from the skin increases, and this structure provides diminished resistance to gas flow. In the interior of the hollow fiber membrane, an open cellular matrix appears, which offers minimal resistance to the transport of gas, as shown in Figure 3. The walls of the cellular structure are formed by aggregations of the spherical micelles.

EFFECTIVE SEPARATING LAYER

Exposure of these asymmetric hollow fiber membranes to an oxygen plasma is a viable procedure for investigating the membrane structure. With electrodeless radiofrequency excitation, diatomic oxygen molecules are diassociated into free radicals and ions. At low pressures, these moieties are sufficiently long-lived to react at the surface of polymers without altering the bulk properties (5-12). Because oxygen plasma ablation is a low temperature process (5), the internal structure of the sample remains unaltered by thermal effects. Table I gives the helium and nitrogen permeation rates of uncoated polysulfone hollow fiber membranes spun from formylpiperidine/formamide. as a function of oxygen plasma exposure times (50 watts RF, 20 sccm oxygen flow rate). Under these conditions the oxygen plasma ablation rate was calculated to be 11.2±1.1 nm/min. The ablation rate was determined from the weight loss as a function of time of melt pressed polysulfone films treated under identical conditions. Duplicate samples were etched at each of the oxygen plasma exposure times prior to the gas flux measurements. The results are illustrated graphically in Figure 4, the changes in helium and nitrogen permeation rates with oxygen plasma exposure time, and Figure 5, the change,

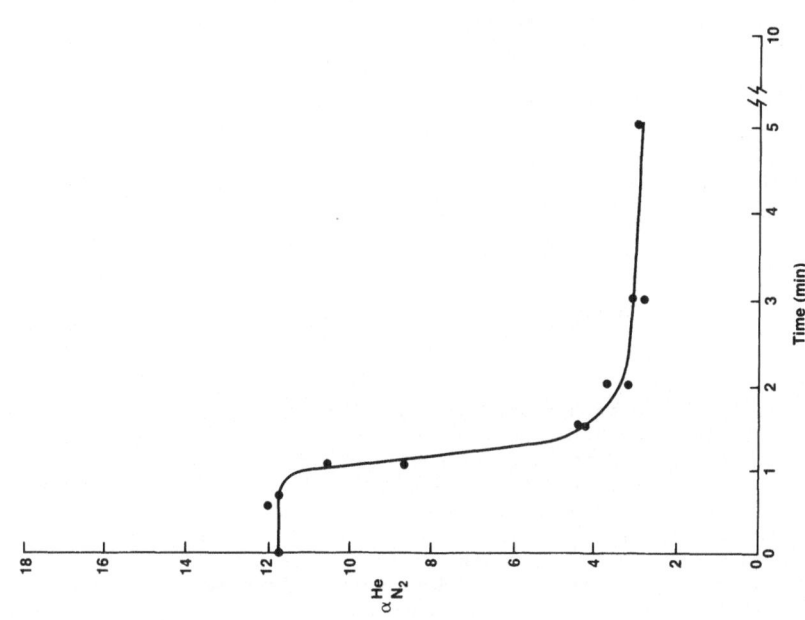

Figure 5. Change in Helium/Nitrogen Selectivity as Function of Oxygen Plasma Exposure Time.

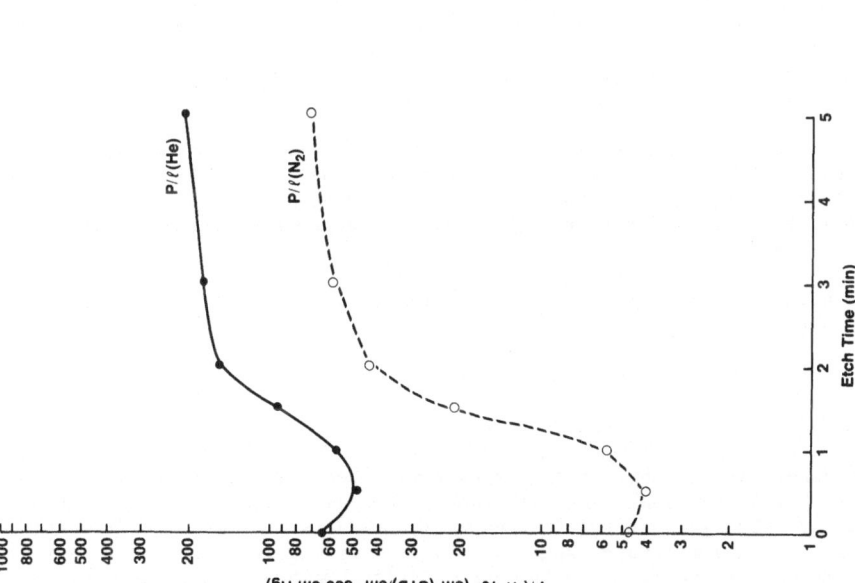

Figure 4. Changes in Helium and Nitrogen Transport Rates as Function of Oxygen Plasma Exposure Time.

21

TABLE II

Performance of Isopentane Treated Polysulfone Hollow
Fibers Spun from Formylpiperidine/Formamide as Function
of Oxygen Plasma Exposure Time

Etch Time (min)	$P/\ell(He)^a$ $\times 10^6$	$P/\ell(N_2)^a$ $\times 10^6$	α
0	60.9	.97	62.8
1	98.0	29.6	4.5
2	221	79.3	2.8
3	171	58.0	3.0
5	329	130	2.5

a. All permeabilities in units of $cm^3(STD)/cm^2\text{-sec-cmHg}$

in selectivity with oxygen plasma exposure time. The helium and nitrogen
permeation rates through the hollow fiber membranes exhibit only minor
changes in the first 30 to 60 seconds of etching followed by a rapid increase
in gas flux rates within the next two minutes of oxygen plasma treatment.
Subsequent oxygen plasma exposure only yields gradual increases in the helium
and nitrogen permeabilities. These results imply a dense effective separating
layer around 6-12nm thick. Beneath the effective separating layer is a
transitional zone, approximately 12-18nm thick, which may contain micropores.
The dramatic tenfold increase in nitrogen permeability with oxygen plasma exposure time is attributed to the exposure of additional micropores with pro-
gressive oxygen plasma ablation. Beyond the transitional region, the flux
through the hollow fiber membrane is controlled by the structure of the
supporting matrix, i.e., porosity size, and pore structure (open versus
closed cells). Examination of the selectivity as a function of oxygen plasma
exposure time, Figure 5, shows that the separation factor of the hollow fiber
membrane exhibits little change during the first 30 to 60 seconds of etching.
Then as abrupt decline in selectivity occurs during the subsequent 30-90
seconds followed by a slow gradual drop in separation factor with additional
plasma ablation, which yields similar conclusions.

The oxygen exposure results indicate that the effective separating layer
in the skin of this asymmetric hollow fiber membrane is probably no more than
18-30nm thick. This is less than 10% of the thickness (500-700nm) of the
microscopically observable skin. These results suggest that the microscopically observable skin is composed of a dense effective separating layer
which is subtended by a dense region containing numerous micropores and
channels with dimensions below the limits of resolution of the scanning
electron microscope.

COATED ASYMMETRIC HOLLOW FIBER MEMBRANES

Membranes for gas separation are extremely sensitive to the existence of
even a relatively few small imperfections. The gas flux, especially that of
the slow gas, through such holes is very high relative to the imperfection-
free areas. Such a membrane exhibits high flux with a low separation factor.
In order to circumvent the need to make a perfect membrane, the membrane is
coated with a thin layer of a highly permeable, nonselective polymer such as a
silicone rubber from an isopentane solution (3,4). The coating plugs the imperfections in the surface of the hollow fiber membrane which sufficiently
reduces the permeability through the pores and defects within the skin to
render permeation through the effective separating layer of polysulfone predominant. Oxygen plasma ablation can also be used to determine the influences
of the coating process on asymmetric hollow fiber membranes. Table II gives

TABLE III

Performance of Sylgard Coated Polysulfone
Hollow Fiber Membranes Coated After Oxygen Plasma Treatment
as Function of Plasma Exposure Time

Etch Time (min)	$P/\ell(He)^a$ $\times 10^6$	$P/\ell(N_2)^a$ $\times 10^6$	α
0	58.1	.52	111
.5	41.8	.36	116
1.0	46.6	.36	131
1.5	46.2	.51	91
2.0	48.7	.58	84
3.0	45.0	.66	68
5.0	50.2	1.22	41
10.0	67.0	10.0	6.7

a. Permeability units in $cm^3(STD)/cm^2$-sec-cmHg

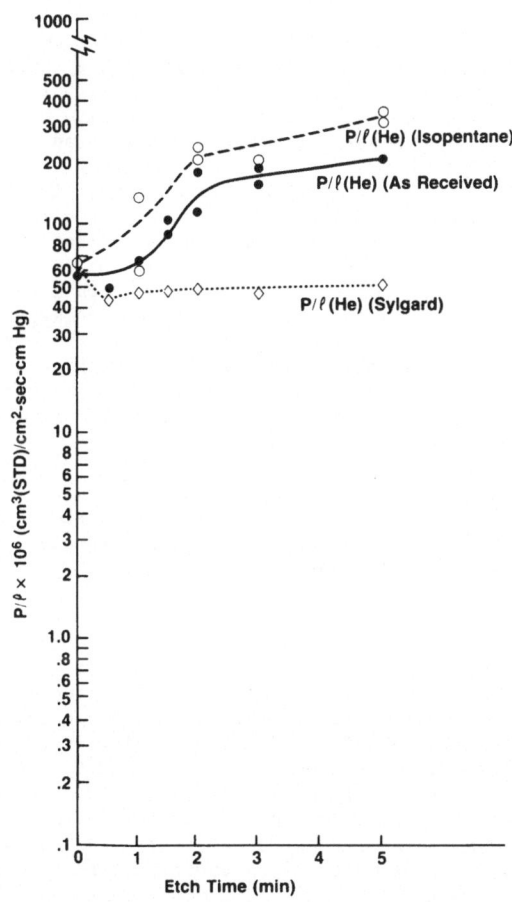

Figure 6. Changes in Helium Transport Rates as a Function of Oxygen
Plasma Exposure Time for As-Received, Isopentane Treated,
and Sylgard Coated After Plasma Treatment Polysulfone Hollow
Fiber Membranes.

the performance of uncoated polysulfone hollow fiber membranes which had been subjected to the coating process with only isopentane solvent present prior to oxygen plasma ablation. Table III lists the results for polysulfone hollow fiber membrane which had been coated with silicone rubber after oxygen plasma ablation. The results given in Tables I-III are illustrated graphically in Figures 6-8. Figure 6 shows the change in the permeability of helium, the fast gas, as a function of oxygen plasma ablation time. The helium permeation rates of the unetched hollow fiber membranes are similar whether the membranes were unexposed, exposed only to isopentane, or coated with silicone rubber. These results indicate that the helium is transported primarily through the polysulfone effective separating layer in the skin of the hollow fiber membrane. Only a small fraction of the helium passes through the imperfections in the skin. During oxygen plasma ablation, the effective separating layer is gradually removed both reducing its thickness and exposing subsurface micropores within the microscopically observable skin. Figure 6 also reveals that the helium permeation rates of the uncoated, isopentane exposed polysulfone hollow fiber membranes exceed that of the unexposed homologs at the same oxygen ablation times. It appears that exposure to solvent decreases the resistance to transport in the interior of the hollow fiber membrane. Possible causes for this phenomenon are that the isopentane densifies material in the matrix surrounding the pores which increases the average pore size, that the isopentane ruptures thin walls of polysulfone separating the pores thus increasing the open cellular structure of the matrix, or that passage of the isopentane reduces the tortuosity in the interior. Coating the polysulfone hollow fiber membranes after oxygen ablation plugs imperfections which were exposed by the oxygen plasma ablation. Because the imperfections in the dense skin are a small fraction of the membrane surface area and transport of the helium through the polysulfone surface layer is rapid, the helium permeability of the coated fiber is relatively constant with oxygen plasma exposure time.

The results for the slow gas nitrogen permeability, illustrated in Figure 7, are more dramatic. Comparison of the results for the samples not subjected to an oxygen plasma reveal a decline in nitrogen permeation rate upon exposure to isopentane. This decline results from the closure of the smaller imperfections in the surface of the hollow fiber membrane from interaction with isopentane. With a reduction in the surface porosity, a larger fraction of the transported nitrogen is that which diffused through the effective separating layer. Coating the hollow fiber membrane with silicone rubber plugs the remaining imperfections reducing the amount of nitrogen transported through these imperfections and causing an additional decline in the measured permeation rate of the hollow fiber membrane.

Oxygen plasma ablation of the isopentane exposed membrane yields samples with higher nitrogen flux rates than those similarly treated but unexposed to isopentane. These results also indicate that the isopentane alters the internal structure of the hollow fiber membrane.

The hollow fiber membranes which were coated after oxygen plasma exposure exhibit an initial decline in nitrogen permeation rate followed by a gradual increase upon additional oxygen plasma exposure. This phenomenon results from the efficiency of coating imperfections with a distribution of sizes. The smaller imperfections are sufficiently small so that the polymer chains of the coating material can cover them but not enter and effectively plug them. The intermediate sized imperfections can be readily plugged by the polymer coating while the largest can be neither plugged nor covered. One would expect that ablation of the surface of the hollow fiber membrane not only removes surface material and reveals subsurface micropores but increases that sizes of the imperfections by removal of the material on their surfaces.

24

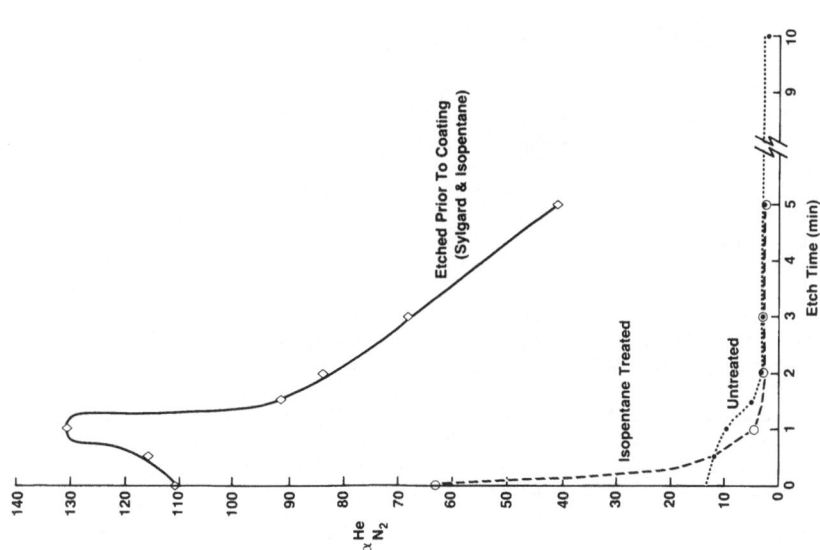

Figure 8.   Changes in the Helium/Nitrogen Separation Factor as a Function of Oxygen Plasma Exposure Time for As-Received, Isopentane Treated, and Slygard Coated After Plasma Treatment Polysulfone Hollow Fiber Membranes.

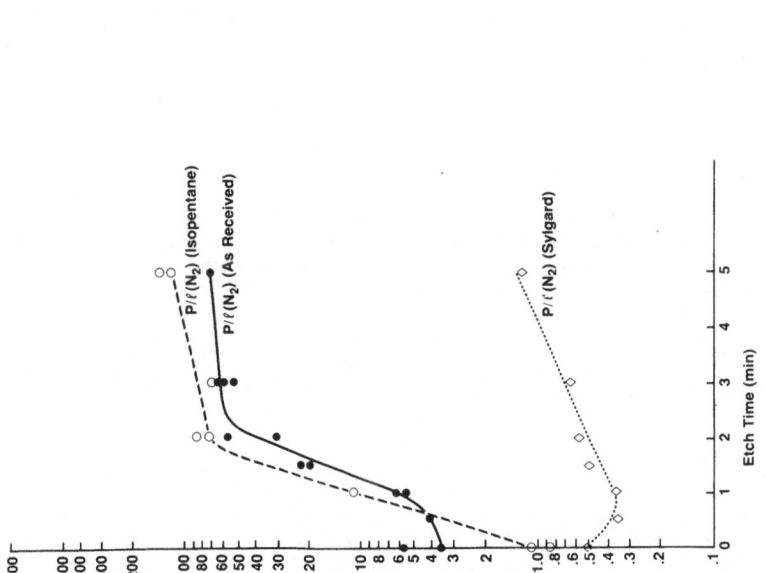

Figure 7.   Changes in Nitrogen Transport Rates as a Function of Oxygen Plasma Exposure Time for As-Received, Isopentane Treated, and Sylgard Coated After Plasma Treatment Polysulfone Hollow Fiber Membranes.

Consequently, if the sizes of the pores were increased by oxygen plasma etching to that more susceptible to efficient coating, then a decline in slow gas permeation rate through the hollow fiber membrane would be expected. However, as oxygen plasma ablation proceeds, both the porosity of the surface would increase due to the exposure of subsurface pores and the average pore size would increase due to removal of material by the oxygen plasma from the walls of the pores. Therefore, an increase in slow gas permeation rate through the coated hollow fiber membrane would be expected.

The effect of isopentane exposure and coating as a function of oxygen plasma exposure time on helium/nitrogen separation factor is shown in Figure 8. Exposure to isopentane is sufficient to increase the selectivity of the uncoated unetched hollow fiber membrane almost fivefold. This increase results from the closure of the smaller imperfections in the surface layer of the membrane from interaction with isopentane. Coating the unetched hollow fiber membranes yields a further increase in selectivity due to the covering or plugging of the remaining imperfections.

Both the isopentane treated and untreated polysulfone hollow fiber membranes exhibit declines in selectivity with oxygen plasma exposure time. This decrease results from the more rapid increase in nitrogen permeability than in helium permeability with oxygen plasma exposure time. After 30 seconds of oxygen plasma exposure, the selectivity of the isopentane treated membrane falls below that of its untreated congener. This implies that the isopentane treatment not only affects the effective separating layer of the hollow fiber membrane to yield the enhanced unetched selectivities but altered the internal structure of the hollow fiber membrane to decrease its resistance to gas transport.

The separation factor of the coated hollow fiber membrane initially increases and decreases with oxygen plasma exposure time. This phenomenon is an artifact of the coating process which limits passage of the slow gas through the surface imperfections of the hollow fiber membrane.

CONCLUSIONS

Asymmetric polysulfone hollow fiber membranes can be prepared by phase inversion spinning solvent/nonsolvent dopes. These asymmetric hollow fiber membranes possess a microscopically observable skin supported by a porous open cellular matrix. Oxygen plasma ablation studies indicate that the effective separating layer of the asymmetric hollow fiber membrane is only a small fraction of the thickness of this microscopically observable skin, which contains pores and channels of dimensions below the limits of resolution of the microscope below the effective separating layer.

Coating the hollow fiber membrane with a nonselective coating reduces the permeability through the imperfections which penetrate the effective separating layer rendering permeation through this layer predominant. The solvent for the coating also alters the structure and morphology of the hollow fiber membrane. The solvent interacts with the hollow fiber membrane to close small imperfections in its surface as well as to reduce the resistance to gas transport in the underlying regions.

REFERENCES

1. S. Loeb and S. Sourirajan, U.S. Patent 3,133,132 (1964).
2. R.E. Kesting, "Synthetic Polymeric Membranes, A Structural Perspective", 2nd Ed., John Wiley and Sons, New York (1985).
3. J.M.S. Henis and M.K. Tripodi, U.S. Patent 4,230,463 (1980).
4. J.M.S. Henis and M.K. Tripodi, J. Membr. Sci., 8, 233 (1981).
5. R.H. Hansen, J.V. Pascale, T. Benedictus, and P.M. Rentzepis, J. Polym. Sci., Part A, 3, 2205 (1965).
6. H. Yasuda, "Plasma Polymerization", Academic Press Inc., New York (1985).
7. E.L. Lawton, J. Appl. Polym. Sci., 18, 1557 (1974).
8. R.H. Hansen, in "Interface Conversion", P. Weiss and G.D. Cheevers, Eds., Elseveir, New York (1969).
9. H. Yasuda, C.E. Lamaze, and K. Sakaoku, J. Appl. Polym. Sci., 17, 137 (1973).
10. G.A. Byrne and K.C. Brown, J. Soc., Dyers Colour., 88, 113 (1972).
11. M. Hudis, J. Appl. Polym. Sci., 11, 1461 (1967).
12. H. Schonhorn and R.H. Hansen, J. Appl. Polym. Sci., 11, 1461 (1967).

COATINGS

Thomas J. Miranda

Whirlpool Corporation

Benton Harbor, MI 49022

INTRODUCTION

The coatings industry has now exceeded the $10 billion mark in sales and approached the billion gallon mark in 1986. The breakdown of coatings use is essentially between three sectors of industry: Trade sales or architectural coatings accounting for about $4.1 billion, OEM or industrial coatings $3.5 billion and specialty coatings, $2.4 billion. Trade sales is the largest sector and the biggest money earner. Industrial sales are led by the transportation industry for cars, trucks, buses and aircraft. This is followed by the container industry, coil coaters, wood finishers and machine and equipment producers. The appliance industry consumes large quantities of coatings including powder, water borne, solvent soluble and coil coatings.

Coatings have made inroads in areas where metallic coatings have previously been used. An example is in auto bumper plating, where chrome plating has largely been replaced with scuff resistant flexible coatings. New two component urethanes are being used on the Ford Taurus and Mercury Sable cars.(1)

This chapter will review some significant applications of industrial coatings.

WATER BORNE COATINGS

Water borne coatings have become a very important factor in both trade sales and in industrial coatings. The acceptance of aqueous coatings has been driven by several factors such as environmental concern for solvent emission reduction, cost and availability of solvents, and reduction in fire hazards. By and large, the greatest impact of water dilutable coatings must be derived from applications in electrocoating.

ELECTROCOATING

Electrocoating, (E-Coat, Elpo, Electrodeposition) of industrial coatings was developed from the pioneering work of George Brewer who recognized the importance of reducing edge corrosion and hidden recess protection in automotive finishes (2). The traditional solvent dilutable flow coat primers when removed from the flow coat stream tended to be redissolved by the evaporating solvent leading to bare spots which became the focal point for rocker panel and inner door corrosion.

Electrocoating provided a means for solving this costly problem. The early electrocoating formulations were derived from the work of Alan Gilchrist of Glidden who prepared maleinized oils which could be rendered water dilutable (3). This was followed by more sophisticated polymers based upon acrylic, epoxy and styrene-allyl alcohol polymers.

These polymers, when neutralized with base, become water dilutable and yield anionic macromolecules which can be deposited on an anode under the influence of an electric current thus:

$$----CH_2-CH--- \ CH-CH--- \quad + \ NH_4OH \quad = \quad ----CH_2-CH--- \ CH-CH---$$

```
    ----CH2-CH--- CH-CH---    + NH4OH  =  ----CH2-CH--- CH-CH---
          |       | |                             |       | |
          R     O=C C=O                           R     O=C C=O
                 \ /                                     |   |
                  O                                     HO  O- NH+
                                                                4

          INSOLUBLE                                       SOLUBLE
```

Anodic electrocoatings provided improved corrosion resistance since, after deposition, the deposited wet film is not washed away by the rinsing water and remains in place in recessed areas. The phenomenon of throwing power, a measure of the coatings ability to coat recessed areas, could be adjusted to obtain adequate coverage of exposed surfaces. In addition, as the film is deposited, its resistance increases dramatically favoring deposition on uncoated metal. Shortly after anodic electrocoating was put into industrial practice, May(4) showed that during the deposition process, there was dissolution of the phosphate metal treatment. This caused soluble ions to diffuse into the depositing film and provided sites for the formation of corrosion cells as water could diffuse into the film driven by the presence of soluble ions.

Another approach to electrocoating would be the deposition of a polymer at the cathode. There were several disadvantages and serious obstacles to this approach since, at the cathode, hydrogen is evolved and would blow away any deposited film:

$$H_3O+ \ + \ e- \quad = \quad 1/2 \ H_2 \ + \ H_2O$$

The development of cationic electrocoating (5) must be one of the milestones in coatings technology and a credit to its inventors who were not discouraged by the advice of a consultant who suggested they drop their research program (6).

The first industrial application of cationic electrocoating was in the appliance industry (7) where the superior properties of cationic electrocoating was demonstrated on air conditioner compressors. This technology was proven in a small (4,000 gallon) tank and, subsequently, a 42,000 gallon tank was installed to coat air conditioner cabinets to

Figure 1. Cationic Electrocoating
of an Air Conditioner Cabinet

replace a two coat system with a highly automated single coat cationic epoxy finish. (Fig. 1)

Following these successes, the automotive industry switched to cationic electrocoating. As a result, improvements in coating performance over anodic systems were quickly demonstrated with salt spray resistance of over 1000 hours exposure.

Early cationic electrocoating systems were deposited in thin films, about 0.4 mils. Recent developments permit coatings thicknesses up to 2.7 mils with significant improvement in performance properties. Thus, as primers, these coatings can perform as porcelain replacements for critical laundry products such as tubs, baskets and tops and lids. With a suitable top coat, salt spray and detergent resistance values up to 1000 hours are achievable. (Fig. 2).

Pure whites are difficult to obtain with cationic epoxy resins, but recently acrylics have been developed which produce excellent whites suitable for appliance applications. The only disadvantage is that the acrylics do not have the corrosion resistance of epoxies.

Acrylic electrocoatings have been applied in the farm implement field where single coat top coats for tractors have been applied with excellent chalk resistance and high gloss.

HIGH SOLIDS COATINGS

For many years, chemists did not concern themselves with the solvent emissions from coatings. This situation changed dramatically in the early sixties when the Los Angeles Pollution Abatement District formulated Rule 66 in an effort to control photosensitive hydrocarbons derived from paint solvents. Coatings were formulated between 20-34% volume solids with flow coating or spray viscosities being adjusted by additional solvent, all of which was discharged to the atmosphere. Rule 66 was followed by more strict legislation greatly reducing the permissible amount of solvent and requiring higher solids coatings to meet set standards. The Environmental Protection Agency regulated certain industries permitting controlled amounts of solvent per gallon. For example, the automotive industry was limited to 3.8 pounds/gallon solvent while the appliance industry was permitted 2.8 pounds/gallon which translated to 62% volume solids.

Thermosetting acrylics were industry's high performance coatings and for appliances were formulated at 34% volume solids. Higher solids resulted in viscosities above the capability of application equipment such as electrostatic spray guns, discs or bells. To achieve compliance, coatings chemists turned to polyesters which have lower solution viscosities than acrylics and have the necessary breakup, flow and leveling characteristics when applied.

To complement this development, manufacturers of application equipment developed high speed turbines and bells (20,000 vs 900 rpm which could handle higher viscosity coatings in record time to satisfy new EPA standards.

Polyesters are cured by coreaction with melamine formaldehyde, urethane or epoxy resins and may be further modified with cellulosics and vinyls. Polyesters did take a good market share from the acrylics, but recent

32

Figure 2. Cationic Electrocoating of a
Washing Machine Cabinet

developments in oligomer chemistry have made acrylics available for
formulating high solids coatings. These low molecular weight, highly
functional prepolymers have low viscosities, cure at relatively low bake
temperatures, have low solvent content and physical properties which .
satisfy end uses.

Higher solids coatings are used in appliance, automotive, coil and container coatings. Because of the higher solids, application problems
unique to these coatings have been encountered. For example, overspray
is generally tacky compared to conventional solids acrylics. Film
buildup with high solids coatings occurs much faster than with conventional solids coatings. As a result, spray booth cleanup is more complicated which makes high efficency transfer spray equipment mandatory
to reduce overspray losses.

COIL COATING

Coil coating was developed in 1935 by Hunter (7) and is a high speed
process for coating large quantities of sheet steel and aluminum. The
process is continuous involving the following steps: uncoiling of a
roll, cleaning and metal treatment, prime coating, baking, top coating,
inspection and recoiling. Line speeds up to 750 feet per minute are
achieved with short bake cycles. This process requires strict performance tolerances in rheological properties to accomodate the high speed
application, flow and leveling and must have good abrasion resistance.

Coil coating has been a boon to the home construction industry and for
large buildings.

Polymer types used in the preparation of coil coatings vary from alkyds
and polyesters to urethanes, vinyls, silicones and fluorocarbons which
span the cost and performance spectrum of architectural requirements.
Recently, the coil coating industry has turned their attention to other
markets and have targeted the appliance industry. Some typical early
applications were to small freezers and range products. Today, coil
coatings are making significant inroads in refrigeration where the
process lends itself to high speed automated assembly, manufacturing
flexibility and replacement of inhouse coating lines. Laundry products
may be slower to accept coil due to the many punched out holes required
in manufacture which form raw edges that are subject to corrosion.

Coil coatings present a number of problems. First, the cost effective-
ness of coil over inhouse painting is still being debated. Second, the
higher cost of scrap and the high amount of rejects, which the customer
must accept, contribute to the overall cost of coil. Coil coatings are
generally smooth and do not have orange peel which is a desirable effect
in some industries, since orange peel does not telegraph surface defects
as much as smooth coatings. Finally, coil coatings cannot be welded
without burning the surface. This problem can be resolved using metal
stitching and structural adhesive bonding.

RADIATION CURED COATINGS

These coatings attracted industrial attention with the early work of
W. Burlant who, while at the Ford Motor Company, was investigating new
methods for curing automotive finishes at lower temperatures (8). His
work led to the curing of coatings on plastic substrates for dashboards

and in developing an experimental line to coat aluminum sheet. Burlant used the then available polyester resins thinned in styrene which required a nitrogen blanket to prevent formation of tacky surfaces.

Radiation cured coatings are those which can be cured under the influence of electromagnetic energy derived from ultraviolet light (UV) or from electron beam sources (EB). Radiation cured coatings are being used for adhesives, coatings and printing inks because they are environmentally acceptable, cure at high speeds at room temperature and are low in energy consumption. These coatings differ from conventional coatings because the solvent becomes part of the final film rather than being evaporated to the atmosphere.

The key to radiation curable coatings is the presence of a conjugated vinyl group, $CH_2=CH$, which readily reacts with the free radicals

$$\underset{\underset{C=O}{|}}{}$$

formed during irradiation. The introduction of multifunctional monomers by a number of chemical companies, Celanese, Rohm and Haas and Union Carbide provided a great impetus to commercialization of these coatings. Some typical monomers include:

- Glycol tri and tetra acrylates or methacrylates; such as ethylene glycol dimethacrylate.
- Acrylated epoxy monomers based on bisphenol A glycidyl ethers
- Acrylated urethane adducts based on toluene diisocyanate or hexamethylene diisocyanate trimer
- Acrylated acrylic oligomers

UV curable coatings are irradiated with sources which produce energies in the 300-430 nm range with energies of from 100-300 watts/sq. in. and are capable of curing coatings up to 10 mils thick. UV coatings require photoinitiators and activators. Such a system might include benzophenone-dimethylethanol amine which interact to produce a radical initiator thus:

$$Ar_2C=O \ + (CH_3)_2N-CH_2-CH_2-OH \longrightarrow [Excited\ state]$$

$$CH_3-N-CH_2-CH_2-OH \ + nM \longrightarrow \ CH_3-N-CH_2-CH_2-OH$$
$$CH_2 \cdot \qquad\qquad\qquad\qquad CH_2-(M)_{n-1}M \cdot$$

(Ar=phenyl)

Electron beam energies are more energetic, 10^5 vs 10^3 e.v., than UV energy and do not require photoinitiators. Also, EB is more penetrating and can cure pigmented systems, although newer UV coatings are capable of curing pigmented systems. In a typical process, the coating, ink or adhesive is applied by spray, knife, roll or curtain coater, then immediately passed under the electron beam or UV source at room temperature. Curing occurs in a matter of seconds. For a review of radiation cured coatings see Ref 9.

Recent developments in radiation curing include coatings for fiber optics and video discs. Early UV coatings were limited to line of sight curing, but recently developed systems based upon carbonium ion cured aliphatic epoxies can and do cure around corners. Early EB accelerators required large shielding facilities, but todays EB units are compact and modular.

Recently, Koleske (10) developed conformal coatings based on aliphatic epoxies which cure rapidly with UV light at low energies and have outstanding arc resistance. These coatings are valuable for circuit board applications and for electronic modules to prevent shorting caused by electrostatic buildup.

POWDER COATING

Powder coatings, which were developed in the fifties in Germany, were applied by a fluid bed process. The powder, an epoxy, is suspended by air blowing through the bottom of a container to form a fluid bed. Parts to be coated are then heated to a high temperature, over 400°F, and immersed into the fluid bed where a coating film is deposited and the residual heat of the part completes flowout and leveling. This process lends itself to large castings and pipe coatings where thick films are possible.

Powder coating is basically a 100% solids coating system. Small amounts of volatiles, up to 8%, may be emitted during the curing cycle depending upon the system. Powders offer a viable method for reducing solvent emissions in the coating process and have recently made significant gains in the Coatings Industry.

Powders are manufactured by melt extruding a polymer, curing agents, pigments and flow control additives onto a cooling conveyer then pulverizing the finished product and grinding to appropriate specifications.

Powders are then applied by fluid bed, electrostatic spray or cloud chambers to parts and then baked to fuse and crosslink the coating. Polymer types used in powder coatings include polyesters, epoxies, vinyls, acrylics and acrylic modified urethanes.

Some early applications of powder coatings in the appliance field include thermoplastic vinyls for dishwasher racks, a practice which is still in widespread use today. Recently, powder coatings have been used in freezer and refrigerator liners, range coatings and more recently as porcelain replacement coatings for laundry applications. For example, washer tops and lids, usually porcelain coated, are now painted with powder coating using polyester powders applied by electrostatic spray. These coatings have excellent hardness, flexibility, corrosion and detergent resistance and are cost effective.

In laundry applications, epoxy powders are being used for coating dryer drums and bulkheads providing a white color with good sales and marketing appeal.

In the automotive industry, the early application of powder coatings met with failure because the expectations and technology were not sufficiently developed at the time. Today powders are being used for wheel coatings and for primer surfacers and exterior trim parts. Laboratory experiments indicate that powders can be used successfully as clear coats over a colored base coat to provide high gloss and depth of image coatings which enhance showroom appearance (11).

RECENT DEVELOPMENTS

A new development in curing of coatings has been the use of a Vapor Cure process in which a urethane coating is prepared with active isocyanate functionality and rather than cure in an oven, the coating is passed into a chamber containing amine vapor. Coating cures are fast, occur at room temperature and offer a new energy efficient method of curing.

Another vapor type curing process is due to 3M where a fluorocarbon vapor bed is maintained and coatings pass into the vapor to absorb heat and cure. The advantage of this system over an oven is that heat transfer is five times more effective.

Another development occured in the preparation of polymers in which a new mechanism, Group Transfer Polymerization, provides a new route to tailor make polymers. This was developed by Webster (12,13) and involves a silyl ketene acetal catalyst in which a monomer is inserted between a trimethylsilyl group hence the name Group Transfer Polymerization. This has been applied to polymerizing methylmethacrylate with good success. At present, the application is in automotive coatings, but this new development suggests a new opportunity for coatings chemists to tailor make polymers.

CRITICAL ISSUES

There are a number of critical issues which will face the chemical industry in the next few years. One issue is the continued pressure being placed on formaldehyde which is used in many industrial finishes as a curing agent. (Formaldehyde is also important in adhesives, fabric and wood finishes.) The key issue is that formaldehyde is a mutagen and could be severely limited in practice or controls could make its use prohibitive. This challenge represents an opportunity for chemists to develop alternatives to formaldehyde curing and those forward looking firms, who plan to remain in business, must develop these alternatives while we have the luxury of time.

Another issue is the continued pressure being placed on halogenated hydrocarbons, particularly chlorofluorocarbons. These compounds are important in aerosols, refrigerant, foams and as solvents. Removal of these from the industrial scene will be a costly and serious event for the coatings industry and will require support from industrial laboratories to challenge theories such as the ozone depletion theory before costly changes are made. Should the theory prove valid, alternates must be in the wings to provide continuity and cost effective coatings materials.

REFERENCES

1. Anon. Chemical Business. October 1986 p. 12
2. Brewer, G.E.F., Journal of Paint Technology, 45, No. 587, 36 (1973)
3. Miranda, T. J., Journal of Coatings Technology, 57, No. 721 22 (1985)
4. May, C. Journal of Paint Technology, 42 No. 552 43 (1971).
5. Bosso, J and M. Wismer, Chemical Engineering, 1971, 78 (13), 114
6. Wismer, M, Journal of Coatings Technology, 58, No. 743, 57 (1986).

7. Gaske, J. Coil Coating, Federation of Societies for Coatings Technology Series. 1987 Edited by D. R. Brezinski and T. J. Miranda

8. Burlant, W. J., D. Green, and J. Taylor, J. Applied Poly. Sci. 1 No. 3, 296 (1959).

9. Costanza, J., R. Silveri, and J. A. Vona, Radiation Cured Coatings, Federation of Societies for Coatings Technology Series. 1986. Edited by D. R. Brezinski and T. J. Miranda

10. Koleske, J. V. and T. M. Austin, J. Coatings Technology, 58 No. 472 47 (1986).

11. Gribble, P. Metal Finishing Feb. 1987 p. 77

12. Webster, O. W., W. R. Hertler, D. Y. Sogah, W. B. Farnham, and T. V. RajanBabu, Polymer Preprints 24, No. 2, 52, (1983).

13. Sogah, D. Y. and O. W. Webster, Polymer Preprints 24, No. 2, 52, (1983).

APPLICATIONS OF POLYMERS IN PACKAGING

Raymond B. Seymour

Department of Polymer Science
University of Southern Mississippi
Hattiesburg, Mississippi 39406

Handmade metal cans which were introduced in France in 1810, were improved and became the standard containers for processed foods a century later. However, the major breakthrough in packaging was the introduction of transparent regenerated cellulose film (cellophane) packaging which was invented by Brandenburg in Switzerland in 1908 and produced in the U.S. in 1924.

Polymeric Films

Cellophane, which was the major packaging film prior to World War II, has been replaced, to a large extent, by polyolefins, polystyrene and polyvinyl chloride (PVC). Less than 10 thousand tons of cellophane is used annually in packaging, but high density polyethylene (HDPE), low density polyethylene (LDPE), polypropylene (PP), polystyrene (PS), and PVC packaging films are consumed in the U.S. at an annual rate of 150 thousand, 1.3 million, 200 thousand, 22 thousand and 92 thousand tons, respectively.

HDPE film is consumed as merchandise bags, tee shirt bags, trash bags and food packaging. LDPE film is also consumed as merchandise bag and tee shirt bags as well as in food packaging, garment bags and shrink wrap for pallets, etc. PP is used in both unoriented film and in oriented film. PS is used as oriented film and PVC film is used as liners, membranes and packaging films. The $50 billion dollar packaging industry is the world's third largest industry in terms of sales and is the world's largest employer.

Film thickness which is kept as thin as practical, is usually less than 125μ. Heavier gauge film, which is used for blister packaging, is classified as plastic sheet. The principal U.S. film producers are DuPont, Eastman Kodak, Hoechst-Celanese, Grace, Hercules, ICI and Mobil. In addition to producing the previously named packaging films, DuPont, Hoechst-Celanese and ICI also produce high strength polyester film, Eastman Kodak produces cellulose triacetate film, and Dow produces vapor resistant polyvinylidene chloride (Saran) and Hoechst-Celanese produces ethylene-vinyl acetate (EVA) copolymer films (1,2).

Packaging film may be produced by extrusion, calendering, solvent casting, tenter frame or blown film techniques. Solvent cast films are first produced as high solid gels via coagulation followed by additional evaporation of solvents. Films are also produced on calenders consisting of 4 rolls with 3 nips. The thickness of the sheet is determined by the metering nip in the final rolls.

Since it permits film orientation, the tenter frame process which allows orientation in the machine direction (MD) and in the transverse direction (TD) is widely used. Film properties of the biaxially oriented film can be enhanced by tensilizing or machine drawing (MD) of the film between rollers.

Simultaneous biaxial orientation direction from the melt is accomplished by the inflation of a quenched blown film. Superior orientation is obtained by reheating the tube and orienting it by additional blowing.

The extrusion process may be used to coextrude multiple films in which the inner films may be recycled polymer. Melt casting and quenching are the initial steps in the production of LDPE film for bread bags, pallet wraps and biaxially oriented film (3,4).

Thermoformed Sheet

About 500 thousand tons of extruded PP, PS, and PVC sheets are thermoformed annually for packaging applications. It is customary to use extruder screws with localizing mixing sections, with fluted design, long mixing sections with double wave screws, and controlled melting barriers, such as the Mallefer screw (5).

Extruder barrels operate at pressures up to 70 MPa at temperatures up to 400°C. The extruded length to diameter (L/D) ratings are usually 20/1, 24, or 30/1. The temperature is controlled in zones along the barrel length. The extruder may have single or double screw designs equipped with screen packs and appropriately designed dies (6).

Most PVC packaging is produced by thermoforming extruded sheet into custom blisters. Since PVC has low crystallinity, it does not melt but softens over a wide temperature range and can be readily thermoformed at temperature ranges of 115-135°C or 170-185°C. Thermoformed high impact polystyrene (HIPS) is widely used as solid and foamed sheet containers for dairy products and for egg cartons and fast food containers.

Thermoformed containers may be produced by drape forming a heated plastic sheet over a positive (male) mold. However, female molds provide more uniform and deeper thermoformed containers. There are many simple and complex techniques and several sophisticated forming machines available and the selection is dependent on the number of parts and the production cycles as well as the desired uniformity of thickness.

Bottles

The first practical plastic bottle-like containers were LDPE "squeezable" bottles which were blow molded in the mid 1940's. In the late 1950's, lighter weight stiffer bottles were blow molded from parisons of HDPE which were used as containers for bleach, household chemicals and milk.

Blow molded acrylonitrile barrier resin bottles were produced in the 1970's but they were not approved for use as soft drink containers by the U.S. Food and Drug Administration. However, comparable bottles were stretch blow molded from polyethylene terephthalate (PET) and received USDA approval. These are widely used as soft drink containers throughout the world. Over 700 thousand of PET is used annually in the U.S. for blow molded bottles.

Tanks and Drums

Volkswagen and military fuel tanks have been blow molded from high molecular weight HDPE for several years and this application is now approved for U.S. and Japanese automobile fuel tanks. The permeability of HDPE to gasoline may be reduced by using coextruded parisons or by fluorination of the interior of the tanks.

20 l HDPE pails are being blow molded and inserted in paper board packages. Self supporting 20 l and 100 l drums are now widely used as closed head shipping and open top storage containers for corrosive liquids. Specifications for overseas plastic shipping containers have been established by the International Maritime Dangerous Goods (IMDG) Code. Over $50 billion is invested annually in industrial shipping containers. These containers include 30 billion plastic bottles valued at $3 billion.

Solid Waste

Since there has been too little incentive for recycling or proper disposal of used plastic packaging, there is considerable concern about this source of nondegradable litter. Fortunately, plastic containers have a high energy content and can serve as an alternate source of fuel. They are also recyclable but regulations must be established for the separation and collections of discarded plastic packaging materials before the solid waste problem is completely solved.

References

1. J. E. Peters, ed., "Packaging Encyclopedia" Cahner Publishing Company, Boston, MA, 1985.

2. M. Bakker, ed., The Wiley Encyclopedia of Packaging Technology, John Wiley and Sons, New York, NY, 1986.

3. J. Sroston, ed., "Plastic Films", Longmans Inc., New York, NY, 1983.

4. R. A. Elden, and A. D. Swan, "Calendering of Plastics", American Elsevier Company, Inc., New York, NY, 1971.

5. C. Mallefer, Mod Plast 40 1 32 (1963).

6. G. Schenkel, "Plastics Extrusion and Technology" American Elsevier Publishing Company, New York, NY, 1983.

7. F. G. Martelli, "Twin Screw Extruders", Van Nostrand-Reinhold, New York, NY, 1983.

HIGH PERFORMANCE POLYMERS

Raymond B. Seymour

Department of Polymer Science
University of Southern Mississippi
Hattiesburg, Mississippi 39406

Mankind's progress has already been dependent on the applications of high performance polymers. Those polymers, such as proteinaceous tendons, silk and sisal fibers, and wood are not usually classified as high performance engineering polymers, but these naturally occurring products have high tensile strengths and durability and continue to be essential for our survival (1).

The pioneer man-made polymers, such as celluloid, rayon and phenolic resins were not particularly superior to the naturally occurring polymers but they were readily available and they and their successor, such as polystyrene, polyolefins and polyvinyl chloride (PVC) became the standards by which other materials of construction were compared.

In spite of their utility and large volume production, these general purpose thermoplastics were not useful for engineering applications. Nevertheless, superior polymers were developed which met the criteria for engineering thermoplastics, i.e., they could be used for engineering design, such as gears and structural members and maintained dimensional stability at temperatures above 100°C and below 0°C.

Nylons

Its use as a strong fiber rather than a high performance thermoplastic took precedence, but Nylon-66 which was developed by Carothers in the 1930's, became the first and most widely used engineering plastic (2). Nylon-66 is a condensation polymer of a 6 carbon dicarboxylic acid (adipic acid) and a 6 carbon diamine hexamethylenediamine. The methylene groups (CH_2) between the amide groups ($CONH_2$) provide flexibility to the polymer chain and the amide groups serve as stiffening groups.

Both Nylon-66 and Nylon-6, a condensation polymer of ξ-aminocaprolactam are characterized by high impact resistance, good abrasion resistance and resistance to many corrosives.

Injected molded nylon is used as gears and other automobile components. Thrust washers and automobile oil filter tubes and housings are obtained by injection molding fiberglass-reinforced nylon. High impact supertough, amorphous nylon is being used for power tool housings and radiator fans. Over 600 thousand tons of non-fibrous nylons are used annually, worldwide (3).

Polyesters

Carothers produced aliphatic polyesters, which were unsuitable for use as commercial fibers or high performance resins. However, Whinfield and Dickson made condensation polymers of the aromatic dicarboxylic acid (terephthalic acid) and ethylene glycol and produced polyesters (PET) which have become the world's most widely used synthetic fibers (4).

PET was blow molded to produce soft drink bottles in the mid-1970's. The carbon dioxide barrier properties of PET bottles were inferior to those of the acrylonitrile barrier resin container but the latter was not approved by the U.S. Food and Drug Administration (USDA) because of the possible toxicity of residual acrylonitrile.

The phenylene groups in PET provide stiffness and the 2 methylene groups in the chain provide a small degree of flexibility. However, polybutylene terephthalate (PBT) is preferred for applications where greater flexibility is essential. Nucleating agents are added to PET to enhance its rate of crystallization. The heat deflection of PET is increased from $160^{\circ}C$ to 224° by the addition of 30 percent fiber glass.

The annual worldwide consumption of these terephthalic resins is about 1 million tons. In addition to its use in bottles, PET is used for many automotive parts such as distributor caps, fender extensions and housings, home appliances, plumbing components and sports equipment.

The resistance to high temperature of polyester resins is enhanced by the condensation of aromatic diols, such as bis-phenol A with the terephthalic acid. These polyarylates have been used for automotive lenses, and electrical components. About 1000 tons of polyarylates are produced annually worldwide.

Polycarbonate resins, which are condensation products of phosgene and bis-phenol A are also aromatic polyesters. The transparent, amorphous polymers are characterized by high heat distortion temperatures ($150^{\circ}C$) and high impact resistance (500 J/m). These polycarbonates are used for glazing, signs, automobile dashboards and electrical components. Over 300 thousand tons of these plastics are used annually worldwide (5).

Another type of polyester is the polyester carbonate resin which consists of phthalate repeat units along with bis-phenol A polycarbonate repeat units in the polymer chain. These engineering resins, which are produced at an annual rate of 100 tons, have a higher heat distortion temperature and better hydrolytic stability than polycarbonate resins (6).

Acetal Resins

Polymeric residues that are deposited in formalin bottles, are thermally unstable. However, these poly (methylene oxides) or acetal resins can be stabilized by end capping of the anionic polymer of formaldehyde with an acyl or alkyl end group. This white, translucent, crystalline polymer has a heat temperature distortion of 136°C and a notched izod impact resistance of 75 J/m.

Thermally stable acetals are also produced by the cationic copolymerization of trioxane and ethylene oxide. The acetal copolymer has a lower heat distortion temperature (110°C) than the homopolymer. These acetal resins which are produced worldwide at an annual rate of about 200 thousand tons are injection molded to produce valves, connectors and plumbing fixtures (7).

Polyphenylene Oxide

Polyphenylene oxide, which is produced by the copper-catalyzed oxidative coupling of 2,6-dimethylphenol is difficult to process but blends of this polymer (PPO) and polystyrene are readily molded (8). Other polymer blends are discussed in a separate chapter.

PPO blends have a heat distortion temperature of 135°C and a notched impact resistance of 120 J/m. These engineering resins are produced at an annual rate of 125 thousand tons worldwide. They are used as unfilled, glass filled, and foamed resins for desalinization, filter frames, medical devices and business machine housings.

Polyphenylene Sulfide

Polyphenylene sulfide is a crystalline polymer with heat distortion temperatures of 111°C and 241°C for the unfilled and 40 percent fiberglass-reinforced resins, respectively. The annual worldwide consumption of this brown-black engineering resin is about 10 thousand tons. It is used in many electrical and electronic applications (9).

Polysulfones

There are several commercial aromatic polysulfone engineering resins, in which the wulfone group (SO_2) serves as the stiffening group and ether (O) or methylene groups serve as the flexibilizing constituents of the polymer chain (cH_2). These amber-yellow transparent amorphous resins have high heat distortion temperatures (174-274°C), good solvent resistance and notched izod impact resistance of 64-200 J/m.

The annual worldwide sales of polysulfone engineering resins is about 10 thousand tons. These products are used for appliances, microwave oven cookware and electrical components (10).

Phenyl Ether Ether Ketone

Phenyl ether ether ketone (PEEK) resins are produced by the displacement reaction of 4,4'-difluorodiphenyl ketone by potassium hydroquinolate. This gray, crystalline polymer contains carbonyl (CO) stiffening groups and ether flexibilizing groups in the polymer chain.

The heat distortion temperature of 165°C is increased to 280°C by the addition of 30% fiber glass or graphite. The annual worldwide production of PEEK is about 500 tons. It is used widely in aerospace and military applications.

Polyimides

Polyimides which have been available for a few decades are produced at an annual rate of 400 tons. The original polyimides were produced in 1908 by Bogert and Renshaw (11) by the thermal dehydration of 4-aminophthalic anhydride. Most of the modern syntheses involve a precursor polyamic acid which is subsequently imidized (12).

The use of polyimide type resins in aerospace and aeronautical applications has been enhanced by processing improvements resulting from modifications, such as polyamide-imide resins (PAI) which are condensation products of trimellitic trichloride and methylenedianiline. PAI is a gray amorphous injection moldable polymer with a heat distortion temperature of 274°C. It is produced, worldwide, at an annual rate of 90 tons. PAI is used for gears and combustion motor parts.

Injection moldable polyetherimide (PEI) amber, amorphous resins have a heat distortion temperature of 200°C. PEI, which is resistant to solvents and corrosives, is used for memory disks, gears, high temperature switches, and sterilizable surgical appliances.

Conclusions

There are now many engineering resins available for a host of unique applications. The present knowledge of structure-property relationships and the use of blends will result in many new products which can solve other application problems.

References

1. H. F. Mark, Chapter 1 in "High Performance Polymers: Their Origin and Development." R. B. Seymour and G. S. Kirshenbaum,eds., Elsevier Science Publishing Company, New York, NY, 1986.

2. H. F. Mark and G. S. Whitby, eds., "Collected Papers of W. H. Carothers", Interscience Publishing Company, New York, NY, 1940.

3. M. I. Kohan, editor, "Nylon Plastics", Wiley Interscience, New York, NY, 1973.

4. V. V. Korshak and S. V. Vinogradova, "Polyesters", Pergamon Press, New York, NY, 1965.

5. R. O. Carhard, Chapter 3 in "Engineering Thermoplastics: Properties and Applications", Marcel Dekker, Inc., New York, NY, 1985.

6. A. Bettleheim Plast Technol 31 20, 22 (1985).

7. R. B. Seymour, "Polymers for Engineering Applications" American Society for Metals, Metals Park, OH, 1987.

8. A. L. Hay, Chapter 21 in "High Performance Polymers: Their Origin and Development", R. B. Seymour and G. S. Kirshenbaum, editors, Elsevier Science Press, New York, NY, 1986.

9. J. Frados, Plastic Focus 16 (49) 1,2 (1985).

10. E. Helmes, et al, "Chemical Economics Handbook", SRI International, Menlo Park, CA 1984.

11. M. T. Bogert and R. R. Renshaw, J. Am. Chem. Soc. 30 1140 (1908).

12. J. J. King and B. H. Lee, Chapter 19 in "High Performance Polymers: Their Origin and Development", R. B. Seymour, G. S. Kirshenbaum, editors, Elsevier Science Press, New York, NY, 1986.

PAS-2 HIGH PERFORMANCE PREPREG AND COMPOSITES

M. R. Lindstrom and R. W. Campbell

Phillips Petroleum Company
Bartlesville, Oklahoma 74004

INTRODUCTION

In 1983, developmental polyarylene sulfide composites were introduced into the marketplace. The matrix for these thermoplastic composites was the semicrystalline polymer, polyphenylene sulfide (PPS). The PPS-based composites exhibit excellent chemical and solvent resistance and good thermal stability. However, PPS unidirectional composites retain only about 30% of their room temperature strength at 350°F.[1] Therefore, use of PPS composites as structural materials at elevated temperatures is limited.

In 1985, J. E. O'Connor reported on the development of a carbon fiber composite based on the semi-crystalline polymer, PAS-1.[2] More recently, a high temperature, amorphous polyarylene sulfide polymer, PAS-2, has been combined with unidirectional carbon fibers to form a high temperature composite. Properties of neat PAS-2 and PPS resins are compared in Table I. As can be seen, the glass transition temperature, Tg, of PAS-2 (215°C) is about 130°C greater than that of PPS. It accounts for the major difference in the elevated temperature performance of these two materials.

PAS-2 Polymer

PAS-2 is a high Tg resin with exceptional chemical resistance for an amorphous polymer as shown in Table II. With the exception of some aromatic oxygenated hydrocarbons and amines, PAS-2 shows no decrease in tensile properties after 24 hours of exposure to a variety of solvents. The reason for this exceptional solvent resistance, when compared to other amorphous resins such as polysulfone, is believed to be due to "pseudocrystallinity" or strong association of the PAS-2 molecules.

In addition to having good chemical resistance, PAS-2 resin is also characterized as having good toughness and impact resistance. Good toughness is indicated by the presence of a tensile yield coupled with good elongation. Impact resistance, as indicated by unnotched Izod, is over twice that of PPS. The combination of a high Tg, good chemical resistance, good toughness and high elongation make PAS-2 an excellent matrix resin for high temperature, high performance composites.

TABLE I

Comparison of Neat Resin Properties of PPS and PAS-2

Property	PPS	PAS-2
Density, g/cc	1.36	1.40
Tensile Strength, psi	11,400	14,500
Elongation, %	2-20[1]	8
Flexural Strength, psi	21,300	25,700
Flexural Modulus, psi	490,000	460,000
Izod Impact, ft lb/in		
Notched	0.4	0.8
Unnotched	10.8	25.2
Oxygen Index, %	44	46
Glass Transition Temp., °C	85	215
Crystalline Melting Temp., °C	285	none

[1]Depending upon crystallinity.

TABLE II

Chemical Resistance of Amorphous Resins

(% Property Retention After 24 Hours at 93°C)

Solvent	PAS-2	Polysulfone	Polycarbonate
Conc. HCl	90	100	0
30% NaOH	102	100	7
10% FeCl$_3$	100	100	100
Water	97	100	100
Acetic Acid	102	91	67
n-Butyl Alcohol	130	100	94
2-Ethoxy Ethanol	123	0	78
Pyridine	19	0	0
n-Butyl Amine	96	0	0
Methyl Ethyl Ketone	45	0	0
Ethyl Acetate	116	0	0
Tetrahydrofuran	38	0	0
Cyclohexane	112	99	75
Toluene	101	0	0
m-Cresol	0	0	0

Prepreg Production

Continuous carbon fiber prepreg is prepared using a proprietary prepregging process which has been shown to be effective for prepregging a variety of resins.[1,2] The prepreg produced from this process results is a composite material with good fiber impregnation and dispersion. To date, PAS-2 has been prepregged with AS-4 carbon fibers from Hercules. Future plans include prepreg production using other continuous fibers such as glass, Kevlar and ceramic fibers as well as preparing a fabric prepreg.

As with any thermoplastic, PAS-2 prepreg has the advantage of having unlimited shelf life with no special storage conditions. Further, laminates can be readily prepared from the prepreg by spot welding several plies together. Thus various lay-ups can be accomplished even though the prepreg exhibits no tack.

Laminate Production

PAS-2/carbon fiber laminates having various fiber orientations have been prepared by placing the appropriate stacked laminates between the platens of an electrically heated press. Laminates have been prepared using either picture frame or positive pressure molds or by simply pressing the plies together between two flat metal plates. Once fully consolidated, the laminates can be cooled slowly in the press, or they can be transferred to a cold press and cooled rapidly. Vacuum bagging and conventional autoclaving have not been evaluated with PAS-2, but no particular problems are anticipated for these techniques.

Composite Properties

Mechanical properties of PAS-2/carbon fiber laminates were determined using either standard ASTM procedures or procedures suppied by various end users. Table III shows a comparison of the room temperature properties of PAS-2/carbon fiber composites with PPS/carbon fiber composites. As can be seen, properties of the two composites are quite comparable.

Table IV compares the 350°F properties of PAS-2/carbon fiber composites with PPS/carbon fiber composites. At this temperature, PAS-2 composites retain 80-90% of their strength and essentially 100% of their modulus. PPS composites, on the other hand, retain only about 30% of their room temperature strength at 350°F. Consequently, PAS-2/carbon fiber composites have utility in structural applications at 350°F where PPS composites do not. This significant increase in elevated temperature property retention is due to the 130°C higher Tg of PAS-2 as compared to PPS.

Applications

The impetus for developing high temperature composites has taken direction from the Air Force and the aerospace industry. In recent years, the need for structural composites for applications approaching 350°F have been identified, and material suppliers are working to develop materials to fill these needs.

Historically, such materials have been limited to thermosetting resin systems as high temperature thermoplastics had not been evaluated for these applications. More recently, thermoplastic resins have been developed which are reported to retain strength up to 700°F, and these resins, coupled with the programs of the Air Force to thoroughly evaluate thermoplastic composites, have resulted in a flurry of activity in the area of high temperature thermoplastic resins.

As indicated above, PAS-2 composites retain in excess of 80% of their room temperature strength at 350°F, and essentially 100% of their modulus. Therefore,

TABLE III

PAS-2 and PPS Room Temperature Composite Properties

Unidirectional Laminates/60 Weight % Carbon Fiber

Property	PAS-2	PPS
Longitudinal Tensile Strength, ksi	172	182
Longitudinal Tensile Modulus, msi	19	17
Longitudinal Flexural Strength, ksi	153	198
Longitudinal Flexural Modulus, msi	15	16
Longitudinal Compressive Strength, ksi	75	92
Transverse Tensile Strength, ksi	5.6	5.2

TABLE IV

PAS-2 and PPS 350°F Composite Properties

Unidirectional Laminates/60 Weight % Carbon Fibers

Property	PAS-2		PPS	
	Value	% Ret.	Value	% Ret.
Longitudinal Tensile Strength, ksi	146	85	68	37
Longitudinal Tensile Modulus, msi	19	100	16	94
Longitudinal Flexural Strength, ksi	119	78	51	26
Longitudinal Flexural Modulus, msi	16	107	13	81

TABLE V

PAS-2 Composite Chemical Resistance[1]

	JP-4 Jet Fuel[2]	5606A Hydraulic Fluid[2]	2% NaOH[3]	Methyl Ethyl Ketone[3]	Methylene Chloride/ Phenol[3]
Tensile Strength, ksi					
Control	25	25	25	25	25
Exposed	22	20	24	19	24
% Retention	88	80	96	76	96
Tensile Modulus, msi					
Control	1.8	1.8	1.8	1.8	1.8
Exposed	1.8	2.1	1.9	1.8	1.9
% Retention	100	117	108	100	106

[1]60% AS-4 Carbon fibers, 12 ply ± 45 laminates
[2]Immersion at 180°F for 14 days
[3]Saturated cloth at 75°F for 24 hours

PAS-2 composites are candidate materials for use in high temperature aerospace applications. Since PAS-2 resin is amorphous, the solvent resistance of the composite to aerospace fluids is essential. Table V lists the solvent resistance of PAS-2 composites to a variety of aerospace solvents. As can be seen, PAS-2 composites retain greater than 80% of their strength and essentially 100% of their modulus.

CONCLUSIONS

A new polyarylene sulfide resin, PAS-2, has been prepregged into a unidirectional carbon fiber tape and used to prepare laminates. These laminates have been found to have ambient temperature properties similar to those observed for PPS composites. However, the 350°F composite strength and stiffness of the PAS-2 composite are significantly greater than those of the PPS composites indicating that the PAS-2 composites can be used for structural applications at elevated temperatures. In addition, it has been shown that the PAS-2 resin has exceptional chemical resistance for an amorphous polymer. This added chemical resistance will allow the PAS-2 composite to be used in applications where composites based on other amorphous polymers could not be used.

REFERENCES

1. J. E. O'Connor, W. H. Beever and J. F. Geibel, "A New Polyarylene Sulfide Polymer Prepreg and High Performance Composite", International SAMPE Symposium Proceedings, 31, 1313 (1986).

2. J. E. O'Connor, A. Y. Lou and D. G. Brady, "Polyarylene Sulfide Composites", Proceedings of the American Society for Composites, 1, 21 (1986).

COMPATIBILITY AND PRACTICAL PROPERTIES OF POLYMER BLENDS

Rudolph D. Deanin

Plastics Engineering Department
University of Lowell
Lowell, Massachusetts 01854

HISTORY

The rubber, coatings, adhesives, and thermoset plastics industries have a long history of blending polymers to optimize the balance of properties for specific applications (1-3). The thermoplastics industry approached the concept much more cautiously and pessimistically over the past four decades. The need for improved balance of properties,and the potential ability of polymer blends to satisfy this need, converged in the successful commercial development of high-impact polystyrene, ABS, rigid polyvinyl chloride, high-impact polypropylene, modified polyphenylene oxide, and four families of thermoplastic elastomers, during the 1940's to the 1960's. More recently, with growing activity in theoretical analysis and practical applications, polymer blending has become a major area for vigorous growth in the past several years (4-9). Properties of plastics which have most often been improved by polymer blending include processability, strength, ductility, impact strength, tack/lubricity, abrasion resistance, heat deflection temperature, low-temperature impact/flexibility, flame retardance, permeability, environmental stress-crack resistance, UV/biodegradability, and price.

THEORETICAL MISCIBILITY, PRACTICAL COMPATIBILITY, AND PROPERTIES

For two polymers to be completely miscible, down to the sub-molecular (segmental) level, optimum requirements should be similar polarity, low molecular weight, and hydrogen-bonding or other strong specific intermolecular attraction. When such miscibility occurs, we would expect properties to follow a simple monotonic function, more or less proportional to the ratio of the two polymers in the blend (Figure 1). This is particularly useful for processors who do their own in-house polyblending, because it permits them to inventory a few commodity polymers and simply blend them in different proportions to meet the specific requirements of each product they manufacture. It is also usually useful for improving melt processability,

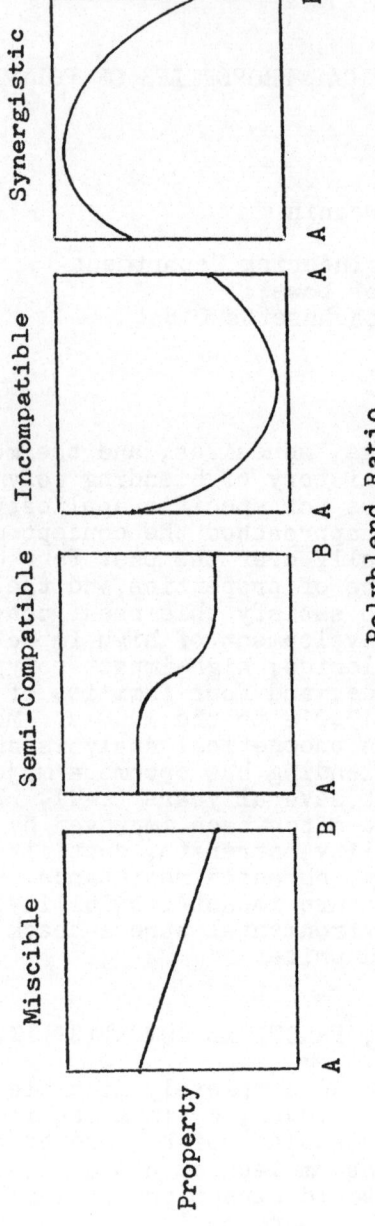

Figure 1. Properties vs. Polymer/Polymer Ratio in a Polyblend

where the added polymer can be chosen to contribute melt fluidity or melt strength as needed.

Most polymer pairs do not meet the above requirements for complete theoretical miscibility. The free energy of mixing is positive, and they tend to separate into two phases. If they are slightly immiscible, each phase will be a solid solution of minor polymer in major polymer, and the phases will separate into tiny sub-microscopic domains, generally well-bonded to each other. Plots of properties vs. the ratio of the two polymers in the blend will be S-shaped, with the polymer present in major amount forming the continuous matrix phase and contributing most toward properties, and an intermediate transition region where there is a phase inversion from one continuous phase to the other. Most of our present successful commercial polyblends have been of this type, with the major polymer forming the continuous matrix phase and retaining most of its useful properties; while the minor polymer forms small discrete domains which contribute synergistically to certain specific properties.

When the polymer pair is much less miscible, phase separation will form larger domains, and the interfacial bonding between them will be weaker. Under stress, the interfaces will fail, and critical properties of polyblends are apt to be poorer than for either of the two polymers in the blend. U-shaped curves of this type are generally taken as a strong indication of immiscibility. In most cases they also signify practical incompatibility, meaning lack of practical utility.

On a few rare and golden occasions, polymer blending can also produce a fourth type of curve for properties vs. polyblend ratio. This is synergistic improvement of properties, beyond what would be predicted from simple monotonic proportionality, and sometimes far exceeding the value for either polymer alone. In most successful commercial polyblends, the same concept can be applied to balance of useful properties, and thus clearly explain the practical success of the polyblend system. While discovery of practical synergistic polyblends is difficult and unpredictable, it is much easier than theoretical explanation of why they are so good. In most cases, such theoretical explanation lies somewhere in the future; and therefore so does our ability to predict them and make them easier to discover and develop.

COMPLICATIONS IN PRACTICE

Until recently the above general treatment appeared to explain the results of polyblend research and commercial practice fairly successfully. With the explosion in polyblend progress during the past few years, however, a great many complications and discrepancies have been accumulating, which indicate the need for much more sophistication, both in basic research and understanding, and also in practical commercialization of polymer blends. Some of the leading problems may be described briefly as follows:

(1) Correlation between polarity and hydrogen-bonding on the one hand, and miscibility and compatibility on the other,

has been far from successful, and it raises the further question whether such correlation should be between whole polymer molecules or between specific groups in the polymer molecules. (2) Composition of many commercial polymers is more or less secret, leaving the researcher unable to explain the different behavior of different members of the same polymer family. (3) The interface in a two-phase system is a very minute area, and analytical techniques are rarely capable of describing its structure and function clearly. (4) It is often assumed that the shape of the property vs. ratio curve is a measure of the separation of the polyblend into separate phases, but the relation to domain size is not clear. (5) It may be assumed that the shape of the curve depends on the sensitivity, of the property being measured, to the size of the domains; thus a gross insensitive property might not distinguish separation into small-size domains, whereas a very fine-scale sensitive property would still see them. (6) There is some indication that simple low-speed tests like tensile modulus and heat deflection temperature are relatively gross properties which ignore separation into micro-phases, whereas complex dynamic tests are more apt to distinguish such fine-scale separation; thus the test method must be chosen both with respect to (a) Theoretical research objectives and (b) Relevance of properties to specific practical applications. (7) When a polymer blend separates into phases, too many researchers assume each phase is a pure polymer, whereas usually each phase is really a solution of the minor polymer in the major polymer; it is very important to remember this in evaluating effects on practical properties. (8) When a polymer crystallizes, addition of a second polymer can modify the crystallization process and thus produce unexpected results which would be difficult to explain without considering this. (9) In reactive systems, simple polyblending is complicated by grafting in thermoplastics; transesterification in polyesters and transamidification in polyamides; and copolymerization in many thermosetting systems; all of these generally increase miscibility and thus have practical effects on compatibility. (10) Current studies are beginning to indicate that processing technique has major effects on polyblend morphology (size and shape of domains), which in turn can have major effects on practical properties. (11) When a specific polymer pair has shown specific effects on properties in a number of studies, we can take these as a reliable basis for both theoretical and practical progress; on the other hand, the first time that such a specific effect is seen, it may require a number of successive studies to determine how reproducible and significant it is. Much current polyblend research is still in this stage.

POLYBLEND RESEARCH AT THE UNIVERSITY OF LOWELL

During the past 12 years, 21 M.S. Plastics thesis research projects at the University of Lowell (10-30) have studied the practical properties of blends of 59 pairs of commercial polymer families (Table 1). Of these, 43 pairs showed useful improvements in properties - either S-shaped curves offering an improved balance of properties, or occasionally outright synergism making the property of the blend superior to either of the polymers in the blend (Table 2).

Table 1. Pairs of Polymer Families Studied

Low-Density Polyethylene/High-Density Polyethylene
Low-Density Polyethylene/Polypropylene
Low-Density Polyethylene/Ionomer
Low-Density Polyethylene/Ethylene-Propylene Rubber
Low-Density Polyethylene/Butyl Rubber
Low-Density Polyethylene/Styrene-Butadiene-Styrene Rubber
High-Density Polyethylene/Polypropylene
High-Density Polyethylene/Ionomer
High-Density Polyethylene/Ethylene-Propylene Rubber
High-Density Polyethylene/Butyl Rubber
High-Density Polyethylene/Styrene-Butadiene-Styrene Rubber
High-Density Polyethylene/Polyurethane
Polypropylene/Ethylene-Propylene Rubber
Polypropylene/Butyl Rubber
Polypropylene/Styrene-Butadiene-Styrene Rubber
Ionomer/Styrene-Butadiene-Styrene Rubber
Ionomer/Polystyrene
Ionomer/ABS
Ionomer/Polyformaldehyde
Ionomer/Polybutylene Terephthalate
Ionomer/Polyurethane
Ionomer/Nylon 6
Ionomer/Nylon 12
Ethylene-Vinyl Acetate/Polyvinyl Chloride
Chlorinated Polyethylene/Polyvinyl Chloride
Styrene-Butadiene Rubber/Polystyrene
Acrylonitrile-Butadiene Rubber/Polyvinyl Chloride
Acrylonitrile-Butadiene Rubber/Cellulose Acetate
Acrylonitrile-Butadiene Rubber/Cellulose Propionate
Acrylonitrile-Butadiene Rubber/Cellulose Butyrate
Polystyrene/Polyurethane
Styrene-Acrylonitrile/Chlorinated Polyvinyl Chloride
Styrene-Acrylonitrile/Cellulose Acetate
Styrene-Acrylonitrile/Cellulose Propionate
Styrene-Acrylonitrile/Cellulose Butyrate
ABS/Polyvinyl Chloride
ABS/Cellulose Acetate
ABS/Cellulose Propionate
ABS/Cellulose Butyrate
ABS/Polycarbonate
ABS/Polyurethane
Polyvinyl Chloride/Polymethyl Methacrylate
Polyvinyl Chloride/Acrylic Polymer
Polyvinyl Chloride/Polycaprolactone
Polyvinyl Chloride/Polyurethane
Chlorinated Polyvinyl Chloride/Acrylic Polymer
Polymethyl Methacrylate/Cellulose Acetate
Polymethyl Methacrylate/Cellulose Propionate
Polymethyl Methacrylate/Cellulose Butyrate
Polymethyl Methacrylate/Polyurethane
Polyformaldehyde/Polyurethane
Cellulose Acetate/Polycaprolactone
Cellulose Propionate/Polycaprolactone
Cellulose Butyrate/Polycaprolactone
Epoxy/Polyurethane
Polycaprolactone/Polyurethane
Polybutylene Terephthalate/Polyurethane
Polyurethane/Nylon 6
Polyurethane/Nylon 11

Table 2. Polymer Pairs Which Gave Useful S-Shaped Curves or
    Synergistic Improvement of Properties (With References)

Low-Density Polyethylene/Propylene: Tensile Strength (16)
Low-Density Polyethylene/Ethylene-Propylene Rubber: Tensile
    Strength, Elongation (23)
Low-Density Polyethylene/Butyl Rubber: Modulus, Tensile
    Strength, Low-Temperature Flexibility (13)
Low-Density Polyethylene/Styrene-Butadiene/Styrene Rubber:
    Modulus, Tensile Strength (25)
High-Density Polyethylene/Ionomer: Tensile Strength, Impact
    Strength, Heat Deflection Temperature (26)
High-Density Polyethylene/Ethylene-Propylene Rubber: Tensile
    Strength, Elongation, Impact Strength, Low-Temperature
    Flexibility (23)
High-Density Polyethylene/Propylene: Modulus, Tensile
    Strength, Heat Deflection Temperature (16,18)
High-Density Polyethylene/Butyl Rubber: Modulus, Tensile
    Strength, Elongation, Low-Temperature Flexibility (13)
High-Density Polyethylene/Styrene-Butadiene-Styrene Rubber:
    Impact Strength (25)
Ionomer/Styrene-Ethylene/Butene/Styrene Rubber: Melt Flow (26)
Ionomer/Polystyrene: Elongation (26)
Ionomer/ABS: Melt Flow (26)
Ionomer/Polyvinyl Chloride: Melt Flow (26)
Ionomer/Polyurethane: Melt Flow, Modulus, Low-Temperature
    Flexibility (26)
Ionomer/Nylon 6: Elongation, Impact Strength (27)
Ionomer/Nylon 12: Melt Flow (26)
Ethylene-Vinyl Acetate/Polyvinyl Chloride: Melt Viscosity,
    Melt Flow, Modulus, Elongation, Impact Strength, Heat
    Deflection Temperature, Low-Temperature Flexibility,
    Thermal Stability (21)
Chlorinated Polyethylene/Polyvinyl Chloride: Melt Flow,
    Modulus, Impact Strength, Heat Deflection Temperature
    (20,29)
Polypropylene/Ethylene-Propylene Rubber: Modulus, Elongation,
    Impact Strength, Heat Deflection Temperature, Low-
    Temperature Flexibility (23)
Polypropylene/Butyl Rubber: Modulus, Tensile Strength, Low-
    Temperature Flexibility (13,15)
Polypropylene/Styrene-Butadiene-Styrene Rubber: Tensile
    Strength, Impact Strength (25)
Styrene-Butadiene Rubber/Polystyrene: Heat Deflection
    Temperature (15)
Acrylonitrile-Butadiene Rubber/Polyvinyl Chloride: Modulus,
    Tensile Strength, Elongation, Impact Strength, Flame
    Retardance (19)
Polystyrene/Polyurethane: Elongation (17)
Styrene-Acrylonitrile/Polyvinyl Chloride: Fusion Rate, Melt
    Viscosity (30)
Styrene-Acrylonitrile/Chlorinated Polyvinyl Chloride: Melt
    Processability (30)
Styrene-Acrylonitrile/Cellulose Butyrate: Modulus, Tensile
    and Flexural Strength (28)
ABS/Polyvinyl Chloride: Melt Flow, Modulus, Impact Strength,
    Heat Deflection Temperature, Flame Retardance(10,14,15,29)
ABS/Cellulose Butyrate: Modulus, Tensile and Flexural Strengths,
    Impact Strength (28)

Table 2(Continued)

ABS/Polycarbonate: Modulus, Tensile Strength, Impact Strength, Heat Deflection Temperature (15,24)
ABS/Polyurethane: Modulus, Elongation (17)
Polyvinyl Chloride/Polymethyl Methacrylate: Modulus, Heat Deflection Temperature (15)
Polyvinyl Chloride/Acrylic Polymers: Fusion Rate, Melt Flow, Impact Strength, Heat Deflection Temperature (11,30)
Polyvinyl Chloride/Polycaprolactone: Melt Flow, Modulus, Tensile Strength, Elongation, Low-Temperature Flexibility, Solvent Resistance, Volatility, Migration (22)
Polyvinyl Chloride/Polyurethane: Elongation (17)
Chlorinated Polyvinyl Chloride/Acrylic Polymer: Melt Processability (30)
Polymethyl Methacrylate/Cellulose Butyrate: Modulus, Tensile and Flexural Strengths (28)
Polymethyl Methacrylate/Polyurethane: Modulus, Elongation (17)
Polyformaldehyde/Polyurethane: Modulus, Tensile Strength, Elongation (17)
Epoxy/Polyurethane: Hardness, Modulus, Flexural Strength, Dielectric Constant, Dissipation Factor (12)
Polycaprolactone/Polyurethane: Elongation (17)
Polybutylene Terephthalate/Polyurethane: Modulus (17)

Some examples: The use of rubbery polymers to improve the impact strength of rigid polyvinyl chloride produced some of the most dramatic demonstrations of synergism; typical examples were ABS (Table 3), acrylic (Table 4), ethylene-vinyl acetate (Table 5), and acrylonitrile-butadiene rubber (Table 6). Similarly the addition of ethylene-propylene rubber to high-density polyethylene (Table 7) and polypropylene (Table 8) again showed the dramatic improvement of impact strength, along with classic S-shaped curves for tensile properties; and addition of butyl rubber to low-density polyethylene similarly showed S-shaped curves for modulus and low-temperature stiffening (Table 9). Where polycarbonate impact strength was very sensitive to thickness, addition of ABS greatly reduced this sensitivity (Table 10). Processability of chlorinated polyvinyl chloride was significantly improved by addition of styrene-acrylonitrile or acrylic processing aid (Table 11). Polycaprolactone served as an effective polymeric plasticizer for polyvinyl chloride, until the excess PCL crystallized out and produced stiffening (Table 12). Copolymerization of epoxy and urethane prepolymers during co-cure apparently produced two-phase systems which gave typical S-shaped hardness and modulus curves (Table 13). And flame-retardance of ABS was dramatically synergized by optimum ratios of polyvinyl chloride and antimony oxide (Table 14).

Some of these have been seen commercially and form the basis of successful commercial products. Others have been seen in more than a single study, suggesting they are dependable enough to be significant. And others have only been seen in a single study, and require further research to determine how reproducible they are, and to what extent they are general for a polymer family or specific to certain members of that family.

The wide range of these observations indicates both the complexity of the problems discussed above, and also the broad scope for the accelerating commercial development of polymer blends as a major sector of the polymer industries in the years to come.

Table 3. PVC/ABS Impact Strength (10)

PVC/ABS	Notched Izod Impact Strength, FPI
100/0	0.59
95/5	0.74
90/10	3.1
75/25	21.3
50/50	11.4
25/75	6.4
10/90	4.1
5/95	3.1
0/100	2.9

Table 4. PVC/Acrylic Impact Modifier (11)

PVC/Acrylic	Notched Izod Impact Strength, FPI
100/0	0.9
90/10	23.6
80/20	27.5
70/30	24.4
60/40	19.7
50/50	15.8
40/60	13.9
30/70	11.4
20/80	8.9
10/90	4.4
0/100	3.5

Table 5. PVC/EVA Impact Strength (21)

% VA in EVA	Effect of PVC/EVA Ratio on Notched Izod Impact Strength, FPI				
	100/0	95/5	90/10	85/15	80/20
28	1.26	19.0	8.7	4.4	1.24
40	1.26	23.2	24.5	11.5	8.1
45	1.26	5.3	18.0	9.0	5.2
52	1.26	23.0	22.5	19.3	18.0
55	1.26	3.7	29.7	27.4	24.0
60	1.26	3.7	5.1	30.5	30.3

Table 6. PVC/NBR Impact Strength (19)

PVC/NBR	% AN	NBR Mooney Viscosity	Notched Izod Impact Strength, FPI
100/0	–	–	0.89
75/25	28	60	17.9
75/25	28	80	19.3
75/25	33	50	1.32
75/25	33	80	18.1
75/25	41	67	1.24
75/25	40	115	27.0

Table 7. HDPE/EPR Polyblends (23)

HDPE/EPR	UTS, PSI	Notched Izod Impact Strength, FPI
100/0	1880	0.54
90/10	1940	13.6
80/20	1740	14.8
70/30	1700	NB
60/40	2000	NB
50/50	1440	NB
40/60	1070	NB
30/70	1090	NB
20/80	850	NB
10/90	660	NB
0/100	530	NB

Table 8. PP/EPR Polyblends (23)

PP/EPR	UE, %	Notched Izod Impact Strength, FPI
100/0	170	2.03
90/10	130	12.7
80/20	150	13.9
70/30	230	NB
60/40	370	NB
50/50	170	NB
40/60	240	NB
30/70	400	NB
20/80	900	NB
10/90	900	NB
0/100	890	NB

Table 9. LDPE/Butyl Rubber Polyblends (13)

LDPE/Butyl	Tensile Modulus, PSI	T(Mod=13K PSI)°C
100/0	24,500	-4
90/10	22,000	-8
80/20	17,700	-16
70/30	15,800	-29
60/40	14,500	-45
50/50	8,850	-60
40/60	3,910	-65
30/70	130	-66
20/80	60	-68
10/90	160	-69
0/100	50	

Table 10. PC/ABS Impact Strength (24)

	Notched Izod Impact Strength, FPI	
PC/ABS	1/8-Inch Sample Thickness	1/4-Inch
100/0	17.8	3.6
90/10	17.8	14.9
75/25	17.2	14.3
50/50	15.2	8.5
25/75	3.2	1.4
10/90	3.4	2.7
0/100	9.8	7.0

Table 11.   CPVC/Polyblend Melt Processability (30)

5 PHR Second Polymer	Melt Flow Gm/10 Min.	Brabender Rheometer 15-Minute Test		
		Fusion Time, Min.	Melt Temp. °C	Melt Torque, M-Gm.
None	0.28	4.7	190	3750
Acrylic	0.48	3.1	180	3490
SAN	0.44	3.8	181	3420

Table 12.   PVC/PCL Modulus (22)

PVC/PCL	Tensile Modulus, PSI
100/0	161,000
95/5	136,000
90/10	133,000
85/15	106,000
80/20	60,600
75/25	18,500
70/30	10,800
65/35	2,450
60/40	940
55/45	12,000
50/50	20,600

Table 13.   Epoxy/Urethane Copolymers (12)

Epoxy/Urethane	Shore D Hardness	Flexural Modulus, PSI
100/0	90	380,000
90/10	88	361,000
80/20	87	338,000
70/30	87	326,000
60/40	84	238,000
50/50	81	165,000
40/60	72	51,200
30/70	64	15,500
20/80	50	2,850
10/90	41	2,460
0/100	39	4,230

Table 14. Flame-Retardance of ABS by PVC/Sb_2O_3 Synergism (14)

ABS	PVC	Sb_2O_3	Burning Rate, In./Min.
100	0	0	1.41
80	0	20	1.36
80	2	18	1.18
80	4	16	1.10
80	6	14	0.80
80	8	12	0.64
80	10	10	0.52
80	12	8	0.25
80	14	6	0.07
80	16	4	0.64
80	18	2	0.82
80	20	0	1.06

REFERENCES

1. M. Morton, "Rubber Technology," Van Nostrand Reinhold, New York, 1973.
2. Fed. Soc. Coatings Tech., "Federation Series on Coatings Technology," FSCT, Philadelphia.
3. I. Skeist, "Handbook of Adhesives," Van Nostrand Reinhold, New York, 1977.
4. B. D. Gesner, Encyc. Polym. Sci. Tech., 10, 694 (1969).
5. R. D. Deanin, Encyc. Polym. Sci. Tech., Suppl. Vol. 2, 458 (1977).
6. J. A. Manson and L. H. Sperling, "Polymer Blends and Composites," Plenum, New York, 1976.
7. D. R. Paul and S. Newman, "Polymer Blends," Academic Press, New York, 1978.
8. O. Olabisi, L. M. Robeson, and M. T. Shaw, "Polymer-Polymer Miscibility," Academic Press, New York, 1979.
9. Mod. Plastics Encyc., 62 (10A), 6,7,34,36,100,154 (1985).
10. R. D. Deanin and C. Moshar, Polym. Preprints, 15 (1), 403 (1974).
11. R. D. Deanin and H. R. Vyas, Coatings & Plastics Preprints, 34 (1), 630 (1974).
12. R. D. Deanin and J. A. Zgrebnak, SPE ANTEC, 20, 654 (1974).
13. R. D. Deanin, R. O. Normandin, and C. P. Kannankeril, Coatings & Plastics Preprints, 35 (1), 259 (1975).
14. R. D. Deanin, R. O. Normandin, and G. J. Ouellette, Coatings & Plastics Preprints, 36 (2), 469 (1976).
15. R. D. Deanin and R. R. Geoffroy, Coatings & Plastics Preprints, 37 (1), 257 (1977).
16. R. D. Deanin and M. F. Sansone, Polym. Preprints, 19 (1), 211 (1978).
17. R. D. Deanin, S. B. Driscoll, and J. T. Krowchun, Jr., Org. Coatings & Plastics Chem., 40, 664 (1979).
18. R. D. Deanin and G. E. D'Isidoro, Org. Coatings & Plastics Chem., 43, 19 (1980).
19. R. D. Deanin and K. B. Sheth, Org. Coatings & Plastics Chem., 43, 23, 27 (1980).
20. R. D. Deanin and M. R. Shah, Org. Coatings & Plastics Chem., 44, 102 (1981).
21. R. D. Deanin and N. A. Shah, Org. Coatings & Plastics Chem., 45, 290 (1981); SPE NATEC, 10/25-27/82, Pg. 286; J. Vinyl Tech., 5 (4), 167 (1983).
22. R. D. Deanin and Z. B. Zhang, Org. Coatings & Appl. Polym. Sci., 48, 799 (1983).
23. R. D. Deanin and S. T. Lim, SPE ANTEC, 29, 222 (1983).
24. R. D. Deanin and C. W. Chu, SPE ANTEC, 31, 941 (1985); J. Elast. Plastics, 18, 42 (1986).
25. R. D. Deanin and Y. S. Chang, SPE ANTEC, 31, 949 (1985); J. Elast. Plastics, 18, 35 (1986).
26. R. D. Deanin and W. F. Liu, Polym. Mater. Sci. Eng., 53, 815 (1985).
27. R. D. Deanin and J. Jherwar, SPE PMAD RETEC, 11/6-7/85, Pg. 24.
28. P. D. Tatarka and R. D. Deanin, SPE ANTEC, 32, 51 (1986).
29. R. D. Deanin and W. Z. L. Chuang, SPE ANTEC, 32, 1239 (1986).
30. R. D. Deanin and J. D. Mast, SPE ANTEC, 32, 1265 (1986).

NEW DIRECTIONS IN INDUSTRIAL POLYSACCHARIDES

Roy L. Whistler
Dept. of Biochemistry
Purdue University
West Lafayette, Indiana 47907

Polysaccharides and their derivatives that are water soluble and produce a high viscosity at low concentrations are commonly called industrial gums. They have a large and growing role in both the food and non- food industry. Some twenty polysaccharides are collected, produced or modified to fill industrial requirements, but the pool of polymers keeps changing as it is modified by availability, application requirements and costs.

One group of industrial polysaccharides undergoing change and even replacement is the group known as exudate gums. They are the ancient gums of commerce used long before biblical times because they were available as encrustations on certain trees, where they could be hand picked, dissolved in water to produce a thick and generally adhesive paste serviceable as food ingredients, for medicinal applications, for printing and sizing fabric and early papers. These were the tree gums of ghatti, karaya and arabic, and the shrub gum, tragacanth. Although these gums still have industrial markets, their usage is diminishing and replacements are actively being sought due largely to lower availability and to increasing costs.

Arabic

Gum arabic, the main exudate gum, has had a very long and fascinating history. It was a commercial article at least 4000 years ago being shipped on Egyptian fleets and described in hieroglyphs as "kami" for use in paints as an adhesive for mineral pigments. It is an exudate of acacia trees of which there are about 500 species distributed over tropical and subtropical areas of Africa, India, Australia, Central America and southwest North America. The most important growing areas are the Sudan where Acacia senegal is produced and Nigeria where several varieties of these Leguminosae occur.

The most important gum arabic yielding area is central Sudan consisting of a belt between 10° to 14° North beginning with western Darfur and extending through the main growing area of Kordorfan with El Obeid as the industrial center on to the Nile River. Gum is obtained from wild or planted groves of acacia that are tapped in the dry season, October, by stripping a 2 foot by 2 inch section of bark and hand collecting the dried tears weighing 20 to 2000g, average 250g, over a 4 to

8 week period. Total production is 60,000-80,000 tons with about half exported to the United States. The government controls production and price. Although the trees are beneficial for the soil and are often planted to prevent desert encroachment, gum harvesting is a laborious and financially unrewarding task. Poor government management reduces gum collection. As with all exudate gum, increase in labor costs can only greatly reduce harvest and increase gum costs.

Ghatti

Ghatti gum exuded from Anogeissus latifolia of the family Comgretaceae is derived mainly from the dry deciduous forests of India and to a lesser extent from Sri Lanka. It occurs in tears of about 2.5 inch diameter but most often in large vermiform masses. Some 1200 tons are collected per year.

Karaya

Karaya gum is an acetylated polysaccharide exudate of Sterculia urens trees growing in dry rocky hills and plateaus of central and northern India, often near the ghatti producing area. Collection is in April to June and again after the monsoons in September. As in ghatti collection, the trees are tapped by removing bark and collecting the dried exudate. About 80% of the annual production of the 5-6 thousand tons is exported to the United States.

Tragacanth

Tragacanth gum is another ancient gum of commerce, being described by Theophrastus in the 3rd century B.C. It is the exudate of the low growing bush, Astragalus, with a long tap root that is tapped for exudation of a long curving string or ribbon of quickly hardening gum called tragacanth, derived from the Greek tragos meaning goat and akantha meaning horn. The leguminous plants are common in Iran, Syria and Turkey; countries not collecting much gum in recent years and especially not much today, although the United States requirement increased to nearly 1,000 tons per year. The price has risen to the extent that companies seek substitutes. One possible relief for tragacanth is to grow high yielding plants in other parts of the world. Thus, Dr. Howard Scott Gentry at the Desert Botanical Gardens, Phoenix, Arizona, who made a collection of the germ plasma of Iranian Astragalus plants just before the Syria-Iranian war, found that many are adaptable to the semiarid regions of southern United States and northern Mexico and in regions where labor costs are relatively low. Plants such as Astragalus echidnaeformis grow well, develop seed and on tapping produce highly acceptable gum in good yield. Even here, the total cost per pound is possibly too high for many applications.

Hemicelluloses as Replacements

An important gigantic source for industrial gums is the hemicelluloses of the plant world where they constitute 25-30% of both annual and perennial plants. Some of these hemicelluloses have properties similar to those exhibited by classical plant exudates. Hemicelluloses are destroyed in present commercial pulping operations to make paper or high alpha-cellulose pulps. They are degraded by the stringent delignification reagents, and will only become available if new delignifications are devised or if special delignification such as that using chlorite or chlorine dioxide become lower in cost. However these problems do not

66

prevent low cost losses from being available in the near future. Initial sources for low cost hemicelluloses are corn fiber from the wet milling industry and the skins of sugar beets and potatoes.

Corn hull gum is obtained from corn fiber which is the membranous covering of the corn kernel. It is the pericarp or more commonly the hull. This structure is built of several layers of dead, thick walled cells that form a tough, impervious covering over the kernel. It comprises 5-6% of the whole kernel and is composed of 49-53% hemicellulose with the remainder mostly cellulose and a small amount of lignin, protein, fat and minerals. This hull would be called bran in the dry milling industry and would contain some remaining unseparated starch. In the wet milling operation, corn grain is steeped to softness, run through attribution mills to tear the grain apart and free the oil-rich germ for centrifugal separation, then further ground between stones to completely rupture the endosperm cells and set free the starch granules that are also separated by centrifugation after screening out the seed coat which is now formed into a fibrous mass by the fine grinding operation. Thus, the seed cost is readily obtained in rather pure form in the normal wet milling process.

Extraction of the corn fiber with alkaline solution dissolves and removes most of the hemicellulose. In one process the corn fiber is extracted with sodium carbonate at pH 10.5-11.5 for 1 hr., the mixture filtered and neutralized to pH 4.0 and the hemicellulose obtained by propanol precipitation or by direct drying. In another process the fiber is extracted with calcium hydroxide to remove a more water soluble portion of the hemicellulose and the filtered solution carbonated with carbon dioxide, followed by filtration removal of the calcium carbonate and the comparatively pure hemicellulose obtained by propanol precipitation after concentration or by roll or spray drying.

Corn fiber hemicellulose is a typical branched cell wall polysaccharide. It consists of a branched backbone of β-$\underline{D}$-xylopyranosyl units with attached α-$\underline{L}$-arabinofuranosyl units, $\underline{D}$- and possibly $\underline{L}$-galactopyranosyl and $\underline{D}$-glucuronosyl units. Its viscosity is slightly higher than that of gum arabic and depending on the manner of isolation equal to or slightly lower than that of gum ghatti. All of its characteristics, thus far examined, suggest that it could be an excellent replacement for one or more of the exudate gums. There seems little doubt that hemicelluloses will soon assume an industrial role of growing importance.

Larch Arabinogalactan

A water extractable hemicellulose from larch, Larix occidentalis, chips has properties that would make this polysaccharide valuable industrially if pricing of a small extraction plant could be justified. For a time the polysaccharide, termed Stractan, was produced by St. Regis Company at their pulp mill at Libby, Montana. Extraction of chips is conducted at ambient temperature counter currently to obtain 8-10% solubles that are then removed on a drum drier. A higher grade free of iron and phenols is produced treating the extraction liquors with magnesium oxide and celite and filtering before roll drying. Chips after extraction move directly into the normal pulping operation where they have added value in being freed of the solubles that consume chemicals and add organic matter to the waste stream. Although the pilot plant operation has been shut down, there is expectation that in todays market the price of larch arabinogalactan would be highly competitive. Supply of larch wood at Libby is 30,000 dry weight tons with another 15,000 tons of saw dust per year.

Modified Hemicelluloses

Hemicelluloses are, of course, subject to modification to alter and improve their properties and this potential extended range of characteristics must surely induce and benefit the early commercial use of this vast source of raw material.

CONDUCTIVE POLYMERS

Raymond B. Seymour

Department of Polymer Science
University of Southern Mississippi
Hattiesburg, Mississippi 39406

Electrical Conductive Polymers

Since many organic polymers are used as thermal and electrical insulators, their use as electrical conductors may be considered as an unusual and unique application. However, polymers with conjugated double bonds do have low ionization potential and high electron affinity and can be readily oxidized or reduced,in the presence of charge transfer agents,which are called acceptor or donor dopants.

Electrical conductivity is also observed in metal-filled polymers and these have been used for electromagnetic interference (EMI) shielding of business machines, etc. More recently, doped films of polyacetylene with good conductivity, have been prepared and used as conductive films (1). The early progress in conductive polymers has been reviewed (2) and will not be discussed in detail in this chapter.

Rechargeable storage batteries have been constructed from two sheets of polyacetylene film separate by an insulating membrane of lithium perchlorate which is placed in a propylene carbonate electrolyte. Doped stretched polyacetylene has a conductivity of $3 \times 10^3 \sigma^{-1} cm^{-1}$. The polyacetylene battery has an open circuit voltage of 3.7v and a current of 0.1 amp cm^{-2}.

Doped poly p-phenylene (3) has a conductivity of $5 \times 10^3 \sigma^{-1} cm^{-1}$ compared to a conductivy of $6 \times 10^8 \sigma^{-1} cm^{-1}$ for copper metal. Polyphenylene sulfide, doped with arsenic pentafluoride and disolved in arsenic trisulfide has a conductivity as high as 50 $\sigma^{-1} cm^{-1}$ (4) which is comparable to doped germanium. Polyaniliane (5), polypyrole and polythiophene are also conducting polymers.

Unfortunately, conductive polymers, such as polyacetylene, are unstable in air, However, doped polythiophene, polypyrrole and polyaniline are stable in air. While both cis and trans polyacetylene are electrically conductive, the trans isomer is more stable and has the better conductivity of the two.

Piezoelectric Polymers

In addition to rechargeable storage batteries, conductive polymers have been proposed for use in EMI shielding and electrical devices such as resistors, capacitors and diodes. Polyvinylidene fluoride is an effective piezoelectric polymer which generates electricity when deformed under stress. When a hot polyvinylidene fluoride film is placed in a DC field, it is polarized and trapped charges remain when the film is cooled in the field (6). These "electrets" are used in microphones, earphones, small loudspeakers, burglar alarms and fire detection devices.

Nylon-11 has about 50 percent the piezoelectricity of polyvinylidene fluoride. These films are used in IR sensitive TV cameras and in submarine detection devices. The piezoelectric properties may be destroyed by heating the films.

Triboelectricification

In contrast to conductive polymers, many non-conductive polymers store electrostatic electricity which attracts dust and may even cause fires. The accumulation of electrical charges by non conductors (triboelectrification) was observed by the ancient Greeks who noted the charge accumulated by rubbed amber. A more quantitative triboelectric series, based on the magnitude of the accumulated charge, was proposed in 1757 (7). The accumulation of electrostatic charges polymers may be prevented by the incorporation of antistats which dissipate the charge.

Photoconductive Polymers

Polymers, such as polyvinylcarbazole, may be photoconductive. It is customary to add a Lewis acid and optical sensitizer or dye to the polymer before electrostatic charging in the dark. This electrophotographic process (Xerography) was invented by Carlson in 1937 but was not used commercially until the late 1940's. In this Xerox process, the image is produced on a polyvinylcarbazole-coated drum, made visible by a toner and then transferred to plain paper and fixed.

Insulating Polymers

As discussed in a separate chapter, foamed polymers are used both as thermal and electrical insulators. The dielectric constant, dissipation factor and dielectric strength are inversely related to the specific gravity of the foamed polymer. Because of their low dissipation factor, these foams are transparent to radar.

Organic polymers, such as extruded PVC, are also used for coating electrical wire and cable. Polytetrafluoroethylene (Teflon) and polyimide are used as a wire coating for service above 200°C. PVC is a preferred insulator for household lamp cord.

Insulation is now a major outlet for polymers and the use of conductive polymers will continue to grow as improved doped polymers are made available. These materials may also be used as electrostatic screens and electroactive elements of probe assemblies.

70

References

1. T. Ito, H. Shirakawa, and S. Ikeda J. Polym. Sci; Polym. Chem. Ed 12 11 (1974).

2. R. B. Seymour, Ed., "Conductive Polymers", Plenum Press, New York, 1981.

3. D. M. Ivory et al, J. Chem. Phys 71 1506 (1979).

4. J. E. Frommer, R. L. Elsenbaumer and R. R. Chance, in "Polymers in Electronic Applications", ACS Symposium Series, 242, American Chemical Society, Washington, DC, 1984.

5. J. A. Pople and S. H. Walmsley, Mol Phys 5 15 (1962).

6. F. E. Karasz, ed., "Dielectric Properties of Polymers", Plenum Press, New York, 1972.

7. V. Shashova J. Polym. Sci (1) 33 65 (1958).

REACTION OF MACRORADICALS

Raymond B. Seymour

Department of Polymer Science
University of Southern Mississippi
Hattiesburg, Mississippi 39406

Introduction

Macroradicals, i.e., electron deficient polymers which may be designated as R·, have played an important role in the progress of animal and plant life since the beginning of time. Of course, the existence of naturally occurring or synthetic macroradicals was not recognized until the 1930's but the existence of macroradicals was also not accepted by many outstanding organic chemists of that era.

Biomacroradicals which are the oldest and most important electron deficient polymers are ubiquitous. They may be advantageous to our well being or detrimental to our health as discussed in the next section.

Biological Reactions

High energy radiation causes cleavage of covalent bonds and thus produces macroradicals in organic polymers, such as DNA, RNA and proteins. These macroradicals may couple to produce crosslinked polymers or they may undergo chain transfer with other organic compounds. In addition to causing wrinkling and lipofuscin (age) spots,the macroradicals may shorten the lifetime of animals unless these free radicals are controlled by the enzymes superoxide dimutase or gluthathione peroxidase. Since the aging process is related to macroradical reactions, the average healthy lifespan can be increased by five or more years by diet selection and the intake of antioxidants,such as vitamins C and E.

Paint

Hubert and Jan van Eyck are credited with the invention of oil painting. The film formation in these oleoresinous paints was catalyzed by the addition of white lead in 400 B.C. The reaction product of unsaturated acids in linseed oil and this basic lead carbonate produced a drier or siccative which promoted the crosslinking or curing of the unsaturated vegetable oils. These film-forming reactions of macroradicals are still in use today for the curing of alkyd and oleoresinous coatings.

Rubber

In 1770, Joseph Priestley unknowingly produced macroradicals when he rubbed caoutchouc (rubber) on paper to erase pencil marks. In 1825, Charles Faraday thermally depropagated rubber macroradicals to produce isoprene which he showed to be a terpene (C_5H_8). Macroradicals were also produced unknowingly in the 1830's when Nathanial Hayward exposed a mixture of rubber and sulfur to sunlight in a "solarization process". These macromolecular reactions were enhanced by Charles Goodyear and Thomas Hancock who crosslinked (vulcanized) rubber by heating.

Hancock also cleaved the covalent bonds in rubber by masticating it in a "pickle" which he invented. A two roll mill and a multiroll calender, invented by Edwin Chaffey in the 1830's, were also used to produce macroradicals by mechanical methods. An intensive mixer, called the Banbury mixer, used to masticate rubber in the early 1900's, reduced the amount of oxygen available for reaction with the isoprene macroradicals produced. In 1904, George Oenslager catalyzed the sulfur-crosslinking of rubber macroradicals by the addition of aniline and its derivatives, such as thiocarbanilide.

In the early 1900'd, it was found that heavy metals, such as manganese, copper, nickel, iron and cobalt, which catalyzed the crosslinking of oleoresinous paints also caused chain scission and thus the formation of macroradicals in rubber. The rate of depropagation of rubber was reduced by Moureau, DuFraisse, Winkleman, Gray and Cadwell, who added condensation products of aromatic amines and aliphatic aldehydes (antioxidants) to natural rubber in the early 1920's. Phenyl-β naphthylamine is now the major antioxidant used in rubber. Hindered phenols and phosphites are used as antioxidants in other polymers. A copolymer produced by milling natural rubber with methyl methacrylate (Hevea Plus) or by polymerizing methyl methacrylate in rubber latex has been produced in Malaysia for several decades.

Protective Coatings

As mentioned previously, controlled crosslinking of macroradicals was used in the paint industry over 2,000 years ago and driers continue to be used today. With the exception of alkyds, most of the replacements for oleoresinous coatings were high molecular weight linear polymers which did not require curing or crosslinking. However, UV radiation and electron beams have been used to cure coatings since the 1960's and the rate of cure has been made almost instantaneous by the addition of photoinitiators, such as benzophenone or benzoin ether. These photoinitiators (sensitizers), are not used for electron beam curing of polymer coatings. These reactions may be prevented by the addition of ultraviolet stabilizers, such as hindered amines (HALS). The crosslinking of asphalt-coated glass by outdoor illumination used in the 1880's, was the first successful photographic process.

Peroxide Curing Agents

Ivan Ostromislensky used benzoyl peroxide as a crosslinking agent for curing rubber in 1915 and comparable initiators have been used for crosslinking ethylene-propylene copolymers (EPM). Dicumyl peroxide has been used to crosslink unsaturated polyesters and silicones. Organic peroxides are also used for the crosslinking of polyethylene, polystyrene and polymethyl methacrylate.

Existence of Macroradicals

In spite of the reactions of macroradicals for thousands of years, the existence of free radicals was not recognized until Moses Gomberg synthesized triphenylmethyl in 1900. While free radical chain polymerization will not be emphasized in this report, it should be noted that such a mechanism was suggested by H. S. Taylor in 1927 and H. Staudinger in 1931. In 1934, Staudinger and W. Heuer observed a dramatic decrease in molecular weight when polystyrene was ball milled or when a solution of this polymer was forced through an orifice.

Mechanochemical Synthesis

It is now recognized that polymer chains can be cleaved to produce macroradicals by mastication, milling, high speed stirring of solutions or alternate freezing and thawing, ultrasonic irradiation, high voltage discharge and by swelling high molecular weight polymers by solvent vapors.

The fate of these mechanochemically produced macroradicals depends on the environment. Recombination and disproportionation occurs in the absence of oxygen, peroxy compounds, chain transfer agents or vinyl monomers. When more than one macromolecule is mechanically cleaved, simultaneously, recombination produces block copolymers.

The macroradical becomes a dead polymer in the presence of chain transfer agents, which are called peptizers in the rubber industry. Both block and graft copolymers are produced in the presence of vinyl monomers. Hevea rubber has been converted to block copolymers by masticating with acrylamide, methacrylic acid, methyl methacrylate, styrene and vinyl pyrrolidone. Graft copolymers of styrene and methyl methacrylate have been produced by milling the corresponding polymers.

Graft Copolymers of Polyvinyl Chloride

Because of the labile character of the chlorine atom, many monomers have been grafted on PVC by using irradiation, cobalt-60, x-rays or linear accelerators. Vinyl monomers, such as methyl methacrylate and acrylonitrile have been added to the trapped free radicals to produce graft copolymers.

Graft copolymers have also been produced by the irradiation of polyvinyl chloride in the presence of vinyl monomers, such as methyl methacrylate, alpha-methylstyrene and mixtures of styrene and acrylonitrile.

Levoprenes, a commercial series of copolymers, have been produced by Farbenfabriken Bayer A.G. by dissolving copolymers and free radical initiators in vinyl chloride monomer. Comparable graft copolymers have also been produced in emulsion systems.

Graft Copolymers of Polyesters

The most widely used commercial graft copolymer was patented in 1941 by Ellis who dissolved prepolymers of ethylene maleate in styrene and added peroxy initiators to produce polystyrene grafts on the prepolymers. Other monomers, such as methyl methacrylate and other unsaturated polyesters, such as acrylic esters of bisphenol A have been used for the production of reinforced plastics, based on polymers of unsaturated polyesters.

Graft Copolymers of Proteinaceous Fibers

In 1944, Speakman patented the grafting of vinyl monomers, such as methyl methacrylate, on wool which had been impregnated with ammonium persulfate. In 1949, Lipson showed that reduced wool produces its own initiator in the form of reduced cystine. Vinyl monomers have also been grafted onto wool by use of irradiation techniques.

In 1953, Negiski and Arai grafted acrylonitrile on silk by using the wool grafting techniques developed by Speakman. In 1965, Imoto grafted methyl methylacrylate on silk and other fibers in the absence of initiators.

Graft Copolymers of Starch

In 1958, Mino and Kaizerman grafted acrylonitrile and other monomers on starch by use of ceric ammonium nitrate dissolved in dilute nitric acid. While styrene did not produce grafts on starch by this technique, grafts with other monomers, such as methyl acrylate and vinyl acetate, have been produced. Grafts with styrene, butadiene and vinylpyrrolidone have been produced by irradiation techniques.

Graft Copolymers with Cellulose

In 1946 and 1951, respectively, Ushakov and Landells grafted vinyl monomers onto cellulose macroradicals using techniques similar to those described for starch in the preceeding paragraph. It is believed that the ceric ion cleaves the anhydroglucose ring between the C_2 and C_3 atoms to produce a macroradical with the single electron on C_2. Cellulose macroradicals and graft copolymers have also been produced by irradiation of cellulose in the presence of vinyl monomers. Acrylonitrile is the most widely used monomer for the production of graft copolymers of cellulose. MTylon-3, which is a soft resilient fiber used in carpet manufacture in USSR is a rayon-acrylonitrile graft copolymer.

Grafting of vinyl monomers to pulp and paper has also been investigated. Graft copolymers of cellulose and acrylic acid ("Super Slurpers") have outstanding water absorptivity. Monomers, such as methyl methacrylate, have also been grafted onto wood, jute and cellulose esters.

Miscellaneous Graft Copolymers

Carlin obtained graft copolymers in 1946 by the polymerization of methyl acrylate in the presence of poly (p-chlorostyrene). In 1952, Rowland grafted ethylene onto polyvinyl acetate. In 1951, Smets grafted vinyl acetate, vinyl chloride and styrene monomers onto polymethyl methacrylate.

Graft copolymers have also been produced by the addition of vinyl monomers to trapped free radicals (microgels) of crosslinked polyacrylonitrile, crosslinked methyl acrylate and crosslinked copolymers of styrene and butadiene (popcorn polymers). Dye acceptor monomers, such as vinylpyridine, have been grafted onto dye-resistant polyacrylonitrile fibers to produce dyable acrylic fibers.

Block Copolymers

Since block copolymers have been considered as a limiting case of graft copolymers and since their properties are similar, graft and block copolymers are often considered as a single branch of polymer science. The first block copolymer synthesized directly from vinyl monomers was a block copolymer of chloroprene formed on methyl methacrylate macroradicals by Bolland and Melville in 1938. In 1956, Melville added methyl methacrylate blocks to vinyl acetate emulsions which had been irradiated by gamma rays.

Block Copolymers of Vinyl Chloride

Nozaki produced block copolymers of monomers, such as acrylic esters and vinyl chloride in the early 1950's. Chimiques Pechiney-St. Gobain patented a process producing block copolymers of vinyl chloride using suspension polymerization techniques and Eilers patented the process for the production of block copolymers of vinyl chloride and methyl methacrylate. Chapiro patented a process for adding monomers, which do not form homopolymers by free radical chain polymerization, such as alpha-methylstyrene to irradiated polyvinyl chloride.

In addition to preparing many block copolymers by the addition of monomers to precipitated macroradicals, Seymour and the coworkers have also produced several block copolymers by addition vinyl monomers to macroradicals in viscous solutions. The viscosity was controlled by the addition of ethylene glycol, glycerol or fumed silica.

Blocked Copolymers by Use of Chain Transfer Agents

Tertiary amines have been used as chain transfer agents in the polymerization of methyl methacrylate. The macroradicals produced by this chain transfer have been copolymerized with other vinyl monomers.

Copolymers With Peroxy End Groups

Initiators, such as phthaloyl peroxide can be used to produce polymers with peroxy end groups. The latter can serve as initiators for block copolymerization with added vinyl monomers.

Block Copolymers of Acrylonitrile

In the 1950's Bamford showed that there was little if any block copolymer formed when styrene was added to occluded macroradicals of acrylonitrile. However, by application of solubility parameter data, Seymour and coworkers were able to produce blocks on acrylonitrile by the addition of charge transfer complexes of styrene and maleic anhydride or styrene and acrylonitrile and showed that styrene would form blocks with these macroradical blocks. These investigators also showed that acrylonitrile readily formed blocks with styrene macroradicals.

A difference in solubility parameter values between polymer and solvent of at least 1.8 H is required for precipitation of macroradicals as they are formed. A difference in solubility parameter values between those of the macroradical and the blocking monomer of less than 3.8 H is required for block copolymerization which must occur in the absence of primary free radicals. These block copolymers are being produced commercially in Japan and Taiwan.

In the early 1900's Shell patented a series of block copolymers produced from macroradicals. Seymour also produced block copolymers by emulsion polymerization techniques in the 1970's. That acrylonitrile and vinyl chloride produced trapped radicals and that stable macroradicals were present in the micells in emulsion polymerization has been known for many years. However, these macroradicals have not been used to produce block copolymers commercially until the 1980's. Information available on solubility parameters and the need to remove primary free radicals should catalyze the production of many new commercial block copolymers.

References

1. Aggarwal, S. L., "Block Copolymers," Plenum Press, New York, 1970.
2. Alliger, G., Sjothien, I. J., "Vulcanization of Elastomers," Reinhold, New York, 1964.
3. Allport, D.C., James, W. H., "Block Copolymers", John Wiley and Sons, New York, 1973.
4. Bamford, C.H., Jenkins, A.D., Nature 176 78 (1955).
5. Bartlett, P.D., Swain, C.G., J. Am. Chem. Soc. 71 1406 (1949).
6. Battaerd, H. A. J., Tregear, G. W., "Graft Copolymers," Interscience, New York, 1967.
7. Bolland, J. L., Melville, H. W., "Proceedings of First Rubber Technol. Conf., London" 1938.
8. Bueche, F., J. Apply. Polym. Sci. 4 101 (1960).
9. Burlant, W. J., Hoffman, A. S., "Block and Graft Copolymers," Reinhold, New York, 1960.
10. Carlin, R. B., Shakespear, N., J. Am. Chem. Soc. 68 876 (1946).
11. Ceresa, R. J., "Block and Graft Copolymers," Buttersworth, Washington, D. C., 1962.
12. Cerisa, R. J., "Block and Graft Copolymerization," Vol. 1 and 2, Wiley-Interscience, New York, 1973, 1976.
13. Compagnon, P., LeBras, J., Rubber Chem. and Technol 20 938 (1947).
14. Ellis, C., U.S. Pat. 2,255,313 (1941).
15. Han, G. E., "Copolymerization" Interscience Publishers, New York, 1964.
16. Hon, D. N. S., "Graft Copolymerization of Lignocellulosic Fibers," ACS Symposium Series 187, Washington, D.C., 1982.
17. Houtz, R. C., Adkins, U., J. Am. Chem. Soc. 55 1609 (1933).
18. Johnson, J. E., et al, "Free Radicals, Aging and Degenerative Diseases," Alan R. Leis, Inc., New York, 1986.
19. Kauzman, W., Eyring, H., J. Am. Chem. Soc. 62 3113 (1940).
20. Molau, G.E., "Colloidal and Morphicological Behavior of Block and Graft Copolymers," Plenum, New York, 1971.
21. Noshay, A., McGrath, J.E., "Block Copolymers," Academic Press, New York, 1977.
22. Oster, G., Phot. Sci. Eng. 4 237 (1960).
23. Ostromislensky, I. I., India Rubber J. 52 470 (1916).
24. Phillips, R. B., et al, Tappi 55 858 (1972).
25. Rowland, J. R., Richards, L. M., J. Polym. Sci. 9 61 (1952).
26. Seymour, R. B., "Block Copolymers," Tankang College Press, Tapei, Taiwan, 1976.
27. Shell Oil, U.S. Pat. 2,666,042 (1952); 3,069,380,1 (1953).
28. Smets, G. Claeson, M. J. Polym. Sci 8 289 (1951)
29. Staudinger, H., Ber. Deut. Ges. 57 1203 (1924).
30. Ushakov, S. N., Fiz—Mat. Nauk 1 35 (1946).
31. Ward, K., Morak, A. J., "Reactions of Cellulose" in "Chemical Reactions of Polymers," Fettes, E. M., Ed., Wiley-Interscience, New York, 1966.
32. Zahran, A. H., Williams, J. L., Stannett, V., J. Appl. Polym.Sci.535 (1980).

FREE RADICAL RING-OPENING POLYMERIZATION OF SPIROTRIENE MONOMERS

William J. Bailey and Jason L. Chou

Department of Chemistry
University of Maryland
College Park, MD 20742

INTRODUCTION

Free radical ring-opening polymerizations are still quite novel in spite of the fact that ionic ring-opening polymerization of heterocyclic compounds, such as ethylene oxide and caprolactam, and Zieglar-Natta ring-opening polymerization of cyclic olefins are well known. In a general research program designed to investigate free radical ring-opening polymerization in our laboratories, it has been found that a wide variety of unsaturated cyclic monomers undergo free radical ring-opening polymerizations including unsaturated spiro ortho esters, spiro ortho carbonates, and cyclic ketene acetals.[1] To investigate the effect of aromatic ring formation on free radical ring-opening polymerization, spirotriene monomers, 3-methylenespiro[5,5]undeca-1,4-diene (1) and 3-methylenespiro[5,6]dodeca-1,4-diene (2) have been synthesized and polymerized.

The earliest example of free radical ring-opening polymerization utilizing aromatization as driving force for ring-opening was studied by Errede,[2] who reported that spiro-di-o-xylylene homopolymerization gave the corresponding poly-o-xylylene:

It was reasoned that homolytic opening of a five-membered ring would cost about 8 kcal/mol of energy (Table 1),[3] while the formation of an aromatic ring would gain the resonance energy about 36 kcal/mol. Hence even with a conservative estimate of the resonance energy of the dienyl free radical, it should be thermodynamically very favorable, if a five- or six-membered ring is opened with the simultaneous formation of an aromatic ring.

From elementary kinetics it is obvious that the extent of ring opening during the polymerization is determined by the ratio of rate of ring opening (k_{iso}) to the rate of direct vinyl propagation without ring opening (k_p). This ratio is influenced by a number of factors including the temperature and concentration of monomer. It is expected that there will be a greater extent of ring opening at the higher temperature or in the more dilute solutions.

1 N = 5
2 N = 6

Table 1.[3]

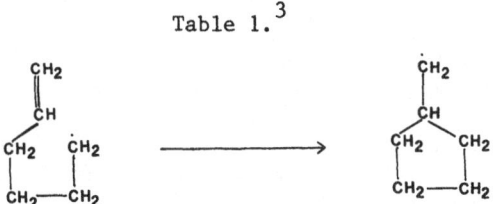

k_{250}, s^{-1}	E, kcal/mol	logA/s^{-1}
1.0×10^5	7.8	10.7

As a result, it has been found that the six–membered spirotriene 1 was polymerized with 67% ring opening at 100°C and 79% ring opening at 130°C. Interestingly, it was shown that solution polymerization of 1 in 75% (by wt) benzene at 130°C gave almost quantitative ring opening. It has also been shown that 1 has a relatively high degree of reactivity as might be expected. For example, copolymerization of 1 with equimolar quantities of styrene at 130°C gave a copolymer consisting of 58% styrene units and 42% of units resulting from 1 of which 92% were ring–opened units.

The seven–membered spirotriene 2 was prepared to investigate the effect of ring size on the free radical ring–opening polymerization. The spirotriene 2 was polymerized with 69% ring opening at 100°C and 100% ring opening at 130°C. While the solution polymerizations of 2 in 80% (by wt) benzene at 100°C gave 93% ring opening and at 130°C gave quantitative ring opening. Copolymerization of 2 with equimolar amounts of styrene at 130°C to give a copolymer consisting of 44% styrene units and 56% of units resulting from 2 of which 100% were ring–opened units.

This ring–opening process makes it possible to introduce the p–phenylene linkage into the backbone of various addition copolymers by a free radical mechanism. The presence of the p–phenylene groups in the copolymers backbone is of particular interest because it would increase the thermal stability of polymers that tend to depolymerize. At high temperatures and high delutions, highly crystalline polymers of the type,

$$\left[\underline{} \bigcirc -(CH_2)_M \right]_N \qquad M = 6, 7$$

can be conveniently prepared.

81

EXPERIMENTAL

3-Methylenespiro[5,5]undeca-1,4-diene (1)

General Procedure.[4] Sodium hydride (1.40 g, 0.035 mol of a 60% dispersion in mineral oil) was washed three times with n-pentane under nitrogen
to remove the mineral oil. Dimethyl sulfoxide (50 mL) was introduced via a
syringe, and the mixture was heated at 70-80°C until the evolution of
hydrogen ceased (ca. 45 min). To the ice bath-cooled dark gray solution,
methyltriphenylphosphonium bromide (11.3 g, 0.032 mol) dissolved in dimethyl
sulfoxide (80 mL) was added. After the resulting dark brown solution was
stirred at room temperature for 30 min, spiro[5,5]-1,4-diene-3-one[5] (5.0 g,
0.031 mol) in 20 mL of warm dimethyl sulfoxide was added. After the addition
was completed, the dark brown mixture was stirred under nitrogen at room
temperature for 3 h, and then poured into water (200 mL). The resulting
mixture was then extracted with n-pentane (5 x 100 mL) and the combined
pentane extracts were washed with 1:1 H_2O-Me_2SO solution (150 mL) and with
water (150 mL), and dried over anhydrous magnesium sulfate. After the
solution had been concentrated by evaporation, the yellow oil was purified
on a silica gel column (100 g) twice by use of petroleum ether (30-60) to
elute the product. The desired product was obtained as a colorless oil in
a 75% yield (3.4 g): ^{1}H NMR (200 MHz, $CDCl_3$) δ 6.15 and 5.90 (AB, J10Hz,
4H, vinyl CH), 4.77 (s, 2H, CH_2=C), 1.70-1.28 (m, 10H, ring CH_2); ^{13}C NMR
(200 MHz, $CDCl_3$) δ 138.46 (C-3), 137.43 (2 endo-vinyl C), 125.48 (2 endo-
vinyl C), 110.86 (exo-methylene C), 38.54 (C-6), 38.29 (C-7 and C-11),
25.91 (C-9), 21.31 (C-8 and C-10); IR ($CHCl_3$) 3080, 3020, 2920, 2840, 1760,
1660, 1580, 1450, 920, 905, 870, 650 cm^{-1}. Anal. Calcd for $C_{12}H_{16}$: C, 89.94;
H, 10.06; mol. wt. 160.1252. Found: C, 89.89; H, 10.11; HRMS, 160.1254.
These results are in agreement with those reported data for 1[6].

Homopolymerization of 3-Methylenespiro[5,5]undeca-1,4-diene (1)

A. In Bulk. In a sealed polymerization tube, 0.5 g (3.12 mmol) of 1
with 15 mg (2 mol%) of benzoyl peroxide were heated at 85°C for 72 h.
After the tube was opened, the mixture was dissolved in $CHCl_3$ and the
resulting solution was then added dropwise into a vigorously stirred
methanol solution. The resulting white precipitate was collected by filtration and dried in vacuo at 80°C overnight to give 0.21 g (42%) of a powdery
white polymer: ^{1}H NMR (200 MHz, $CDCl_3$) δ 7.2-6.8 (br d, aromatic CH),
5.9-5.0 (br AB, nonring-opened vinyl CH), 2.7-2.3 (br s, benzylic CH_2),
2.1-0.5 (br m, ring-opened and nonring-opened CH_2); ^{13}C NMR (200 MHz,

CDCl$_3$) δ 140.18 and 135.63 (2 aromatic q C), 133.81 and 130.38 (2 nonring-opened vinyl $\underline{C}$H), 128.20 and 127,13 (2 aromatic $\underline{C}$H), 41.06 and 38.26 (2 nonring-opened q C), 48.65-21.44 (m, ring-opened and nonring-opened $\underline{C}$H$_2$); IR (KBr) 3010, 2900, 2830, 1510, 1440, 1260, 1100, 1020, 910, 820, 760 cm^{-1}; [η] = 0.235 dL/g (in benzene, at 30°C). Anal. Calcd for (C$_{12}$H$_{16}$)$_n$: C, 89.94; H, 10.06. Found: C, 89.70; H, 10.30.

Similar procedures were applied to the syntheses of other polymers at various temperatures and the identification of their structures.

B. In Solution. In a sealed polymerization tube, 0.5 g (3.12 mmol) of 1 and 1.5 g of dry benzene were heated with 37 mg of t-butyl peroxide at 130°C for 12 h. After the tube was opened, it was found that the obtained white solid polymer chunk was insoluble in common organic solvents including benzene and chloroform at 70°C. The solid polymer was dried in vacuo over-night at 100°C to give 0.5 g (100%) of a white solid polymer: IR (KBr) 3020, 2910, 2840, 1900, 1780, 1605, 1510, 1460, 1360, 1300, 1260, 1200, 1100, 1020, 905, 820, 770, 720 cm^{-1}; ^{1}H NMR (200 MHz, CDCl$_3$/PhOPh) δ 2.52 (br, 4H, CH$_2$-Ph), 1.60 and 1.33 (br, 8H, ring-opened CH$_2$); [η] = 0.233 dL/g (in phenyl ether, at 165°C). X-ray diffraction pattern on an unoriented film showed sharp rings which were the evidence of crystallinity in the obtained polymer; DSC analysis showed T$_m$ = 106°C and T$_g$ = 6°C. Anal. Calcd for (C$_{12}$H$_{16}$)$_n$: C, 89.94; H, 10.06. Found: C, 89.30; H, 10.12.

Copolymerization of 3-Methylenespiro[5,5]undeca-1,4-diene (1) with Styrene

In a sealed polymerization tube, a mixture of 0.5 g (3.12 mmol) of 1, 0.3 g (3.12 mmol) of styrene, and 0.018 g (0.123 mmol, 2 mol%) of t-butyl peroxide was heated at 130°C for 4 h. After the resulting solid chunk was dissolved in CHCl$_3$, the solution was added to methanol to precipitate the copolymer. The collected solid was dried in vacuo at 80°C overnight to give 0.67 g (84%) of a white polymer as a thin film: IR (KBr) 3010, 2900, 2840, 1780, 1600, 1495, 1450, 1370, 1180, 1020, 900, 810, 755, 695 cm^{-1}; ^{1}H NMR (200 MHz, CDCl$_3$) δ 7.30-6.25 (br, m, styrene and ring-opened aromatic H), 5.70-5.20 (AB nonring-opened vinyl H), 2.85-1.70 (br, m, styrene and ring-opened benzylic H), 1.70-0.60 (br, m, aliphatic CH$_2$); ^{13}C NMR (200 MHz, CDCl$_3$) δ 145.51 (styrene q C), 139.99 (ring-opened aromatic q C), 133.84 and 130.54 (nonring-opened vinyl $\underline{C}$H), 129.02, 128.14, 127.85, 127.21, 125.81 (styrene and ring-opened aromatic $\underline{C}$H), 48.03-21.52 (m, aliphatic C). Anal. Calcd for (C$_{12}$H$_{16}$)$_{0.4227}$(C$_8$H$_8$)$_{0.5773}$: C, 91.28; H, 8.72. Found: C, 91.03; H, 8.97; [η] = 0.505 dL/g (in CHCl$_3$, at 30°C).

Cycloheptanecarboxaldehyde (5)

General Procedure.[7] Oxalyl chloride (55.85 g, 0.44 mol) dissolved in
methylene chloride (1000 mL) was placed in a four-necked flask equipped with
a mechanical stirrer, a thermometer, and two pressure-equalizing addition
funnels protected by drying tubes. One addition funnel contained dimethyl
sulfoxide (75.00 g, 0.96 mol) dissolved in methylene chloride (200 mL) and
the other cycloheptanemethanol (50 g, 0.39 mol) dissolved in the same solvent
(200 mL). After the contents of the flask was cooled to -60°C, the dimethyl
sulfoxide solution was added dropwise in ca. 30 min (exothermic). Stirring
was continued at -60°C for 30 min followed by addition of the alcohol solution in ca. 30 min (exothermic). After the reaction mixture was stirred
for 60 min, triethylamine (202.38 g, 2 mol) was added in ca. 30 min with
stirring at -60°C. After the mixture was allowed to warm to room temperature, water (1200 mL) was added. Stirring was continued for ca. 45 min and
then the organic layer was separated. The aqueous phase was extracted with
methylene chloride (800 mL), and the combined organic layers were washed
with a saturated sodium chloride solution (1000 mL), hydrochloric acid (1%,
1000 mL), water (1000 mL), sodium carbonate solution (5%, 1000 mL), and
water (1000 mL). After the organic layer was dried over magnesium sulfate,
the filtrate was concentrated by evaporation to dryness to give 43 g (87%)
of the desired aldehyde 5 with analytical results essentially identical to
those previously reported.[8]

3-Methylenespiro[5,6]dodeca-1,4-diene (2)

The precursor, spiro[5,6]-1,4-diene-3-one (4) was prepared from the
starting material, cycloheptanecarboxaldehyde (5) by use of Kane's procedure[5] in an overall yield of 23%: mp 51-52°C; ^{1}H NMR (200 MHz, CDCl$_3$) δ 7.00
and 6.16 (AB, J10Hz, 4H, vinyl CH), 1.65 (s, 12H, ring CH$_2$); IR (CHCl$_3$)
2990, 2920, 2850, 1660, 1620, 1460, 1400, 1270, 1250, 1170, 865 cm^{-1}. Anal.
Calcd. for C$_{12}$H$_{16}$O: C, 81.77; H, 9.15. Found: C, 81.31; H, 9.34. (Although
the ketone 4 was synthesized by Kane,[5] no analytical data were reported in
the corresponding papers).

Then by use of the Wittig reaction in dimethyl sulfoxide previously
described concerning the synthesis of the monomer 1, the ketone 4 (5.5 g,
0.032 mol) was converted to the desired monomer 2 in a 69% yield: ^{13}C NMR
(200 MHz, CDCl$_3$) δ 138.83 (2 endo-vinyl C), 138.21 (C-3), 124.16 (2 endo-
vinyl C), 110.76 (exomethylene C), 41.03, 30.55, 22.88 (6 ring CH$_2$), 32.42
(C-6); IR (CHCl$_3$) 2990, 2910, 2850, 1760, 1660, 1585, 1455, 870, 650 cm^{-1};

HRMS Calcd. for $(C_{13}H_{18})$: 174.14085. Found: 174.1408507; mp and ^{1}H NMR were in agreement with those reported for 2.[9]

Homopolymerization of 3-Methylenespiro[5,6]dodeca-1,4-diene 2

A. **In Bulk**. In a sealed tube, 0.5 g (2.86 mmol) of 2 and 13.86 mg (2 mol%) of benzoyl peroxide were heated at 100°C for 20 h. After the tube was opened, the mixture was dissolved in $CHCl_3$ and the resulting solution was then added dropwise to a vigrously stirred methanol solution. The resulting white precipitate was collected by filtration and dried in vacuo at 80°C overnight to give 0.4 g (80%) of a light yellow solid polymer: ^{1}H NMR (200 MHz, $CDCl_3$) δ 7.2-6.9 (br d, aromatic CH), 5.58 and 5.20 (AB, non-ring-opened vinyl CH), 2.55 (m, benzylic CH_2), 1.7-0.7 (br m, ring-opened and nonring-opened CH_2); ^{13}C NMR (200 MHz, $CDCl_3$) δ 140.02 and 135.70 (2 aromatic q C), 135.12, 130.63, 128.83, 128.21, 127.13 (aromatic CH and non-ring-opened vinyl CH), 43.07 and 38.19 (2 nonring-opened q C), 48.51-22.67 (m, ring-opened and nonring-opened CH_2); IR (KBr) 3000, 2920, 2850, 1510, 1455, 1260, 1090, 1020, 850, 800, 760 cm^{-1}; [η] = 0.631 dL/g (in benzene, at 30°C). **Anal**. Calcd for $(C_{13}H_{18})_n$: C, 89.59; H, 10.41. Found: C, 89.84; H, 10.16.

Similar procedures were applied to the synthesis of the polymer at 130°C and the identification of its structure.

B. **In Solution**. In a sealed polymerization tube, 0.5 g (2.86 mmol) of 2 and 2 g of dry benzene were heated with 27.72 mg (4 mol%) of benzoyl peroxide at 100°C for 72 h. The resulting mixture was dissolved in $CHCl_3$, and the solution was added dropwise into methanol. The white precipitate was collected and dried in vacuo at 80°C overnight to give 0.25 g (50%) of a light yellow solid polymer: ^{1}H NMR (200 MHz, $CDCl_3$) δ 7.10 (s, 4H, aromatic CH), 2.55 (t, 4H, benzylic CH_2), 1.60 and 1.35 (br s, 10H, ring-opened CH_2); ^{13}C NMR (200 MHz, $CDCl_3$) δ 140.04 (2 q aromatic C), 128.23 (4 aromatic CH), 35.57 (2 benzylic CH_2), 31.52, 29.32 (5 ring-opened CH_2); IR ($CHCl_3$) 3010, 2920, 2850, 1710, 1510, 1450, 1270, 1110, 1020, 815 cm^{-1}; [η] = 0.619 dL/g (in benzene, at 30°C) **Anal**. Cacld. for $(C_{13}H_{18})_n$: C, 89.59; H, 10.41. Found: C, 89.44; H, 10.56.

Copolymerization of 3-Methylenespiro[5,6]dodeca-1,4-diene (2) with Styrene

In a sealed polymerization tube, a mixture of 0.7 g (4.02 mmol) of 2, 0.4 g (3.84 mmol) of styrene, and 22.99 mg (2 mol%) of t-butyl perpxode was heated at 130°C for 4 h. After the resulting solid chunk was dissolved in CHCl$_3$, the resulting solution was added to methanol to precipitate the co-polymer. The collected solid was dried in vacuo at 100°C overnight to give 0.96 g (88%) of white solid copolymer: ^{1}H NMR (200 MHz, CDCl$_3$) δ 7.30–6.30 (br m, styrene and ring-opened aromatic CH), 2.90–1.70 (br m, styrene and ring-opened benzylic H), 1.70–0.70 (br m, aliphatic CH$_2$); ^{13}C NMR (200 MHz, CDCl$_3$) δ 145.54 (styrene q C), 140.03 (ring-opened aromatic q C), 128.96, 128.20, 127.98, 127.77, 125.85, 125.64 (styrene and ring-opened aromatic CH), 47.98–15.27 (m, aliphatic C); IR (neat) 3030, 2930, 2860, 1605, 1515, 1495, 1455, 1030, 755, 695 cm^{-1}. <u>Anal</u>. Calcd. for (C$_{13}$H$_{18}$)$_{0.56}$(C$_8$H$_8$)$_{0.44}$: C, 90.44; H, 9.56. Found: C, 90.75; H, 9.25.

RESULTS AND DISCUSSION

The precursor of <u>1</u>, spiro[5,5]undeca-1,4-diene-3-one (<u>3</u>) was prepared in large scale by use of the Kane's procedure[5] in an over-all yield of 36%:

Then the ketone <u>3</u> was converted to the desired monomer, 3-methylene-spiro[5,5]undeca-1,4-diene (<u>1</u>) via a Wittig reaction[4] in a 75% yield.

The colorless liquid monomer <u>1</u> was stable under nitrogen in a freezer, but at room temperature in air the color gradually changed to brown to give a spontaneous polymerization product with low molecular weight.

Homopolymerizations of <u>1</u> were carried out at different temperatures to study the extent of ring opening as a function of the polymerization temperature. When monomer 1 was polymerized at 100°C over a period of 48 h with 2 mol% of benzoyl peroxide as the initiator, a white solid polymer was isolated after purification. The 200 MHz ^{1}H NMR spectrum of chloroform soluble fraction showed signal at δ 7.0 corresponding to the aromatic protons and at δ 2.5 corresponding to the benzylic protons which were the indication of ring opening and aromatic ring formation. On the other hand, the signals in the range of δ 5.0-6.0 corresponding to the vinyl protons which were the indication of nonring opening units in the polymer (Fig. 1). Thus, the integration of the signals either at δ 7.0 or at 2.5 versus the integration of the signals in the range of δ 5.0-6.0 was used to obtain the ratio of ring opening versus nonring opening (Table 2).

Indeed, in a series of bulk polymerizations, when temperature was increased from 85 to 100, then to 130°C, the extent of ring opening was also increased from 43 to 67, then 79% which was evidence of the fact that higher temperature favored ring opening process.

It has been reported that poly(p-xylylene), which is analogous to our 100% ring-opened polymer is insoluble in many common solvents, and it is only soluble in a few high boiling solvents, such as biphenyl, chloronated biphenyls, and phenyl ether at high temperatures (above 250°C).[10] We also found that the obtained polymers were insoluble in common solvents, such as benzene, chloroform at 70°C, when the extent of ring opening in the polymerization was higher than 61%. When phenyl ether was chosen as a solvent, the polymers were slightly soluble when temperature was above 100°C, and the solubility was a function of the extent of ring opening. Therefore, a special method was used to obtain the NMR spectra: first the polymers were dissolved in phenyl ether at 180°C (about 1 g in 100 mL of solvent); then

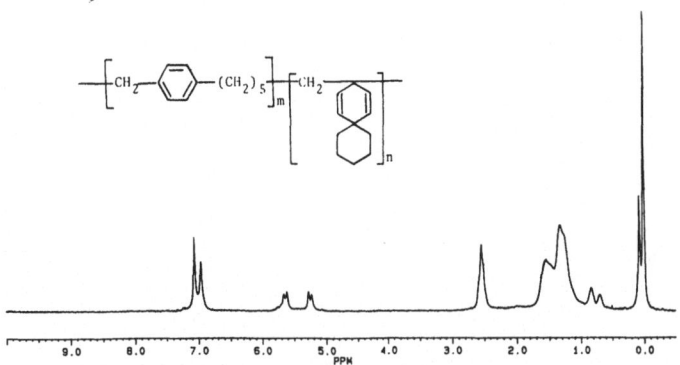

Figure 1.   200 MHz ^{1}H NMR of the polymer obtained from monomer 1

Table 2. Homopolymerization of $\underline{1}$ at Various Temperatures

INITIATOR (mol %)		T(°C)	TIME (h)	POLYMER YIELD(%)	SOLUBILITY	EXTENT OF RING OPENING[a] (mol %)	VISCOSITY [η](dL/g)
Benzoyl peroxide	2	85	72	42	Soluble in CHCl$_3$, C$_6$H$_6$	43	0.235[d]
Benzoyl peroxide	2	100	48	80	33% Soluble in CHCl$_3$ 61 67% Insoluble in common solvents[b] 70	Average 67	0.105[e]
t-Butyl peroxide	4	130	12	100	Insoluble in common solvents[b]	79	0.255[e]
t-Butyl peroxide[c]	8	130	12	100	Insoluble in common solvents[b]	98	0.233[e]

[a]From 200 MHz ^{1}H NMR study
[b]Below 80°C
[c]Solution polymerization (3:1 benzene:monomer XI by wt)
[d]In benzene at 30°C
[e]In phenyl ether at 165°C

the solution was slowly cooled to room temperature to make a supersaturated solution; finally 25% of d-chloroform was added to prepare the NMR sample. In this case, for each polymer the ^{1}H NMR integrations of signals at δ 2.5 and between δ 5.0–6.0 were followed to calculate the extent of ring opening.

To determine the effect of concentration of the monomer during polymerization on the extent of ring opening, monomer $\underline{1}$ was polymerized at 130°C in benzene (1:3 by wt) with 8 mol% of t-butyl peroxide as the initiator for 12 h. The polymer obtained showed a structure different from that of the polymer made at the same temperature in bulk, since the absence of the signals between δ 5.0–6.0 in its ^{1}H NMR spectrum clearly indicated that monomer $\underline{1}$ had undergone almost quantitative ring opening in this solution polymerization, which was evidence that more dilute conditions favored the ring opening reaction.

As predicted, the X-ray diffraction on an unoriented film showed a sharp ring pattern which was an indication that the crystallinity of the solution polymerization polymer was due to its linear and symmetrical structure. A further indication of regular structure of the polymer was shown by a DSC scan which showed a T_m = 106°C (Fig. 2). Presumably, increasing

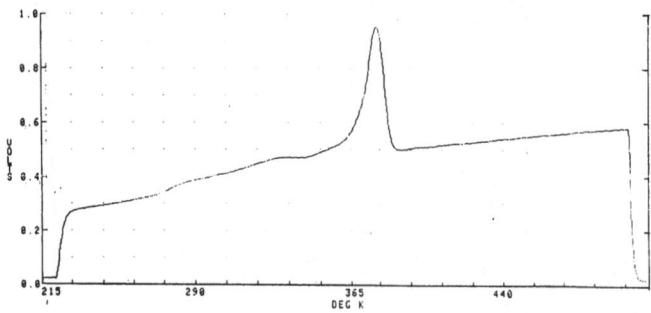

Figure 2. DSC of the polymer made by solution polymerization of $\underline{1}$

the extent of ring opening or decreasing the number of methylene units will increase the melting point, T_m.

In a copolymerization experiment, equal molar amounts of monomer 1 and styrene were polymerized at 130°C in the presence of 2 mol% of t-butyl peroxide as the initiator for 4 h. A solid copolymer obtained in a 84% conversion was isolated after purification by reprecipitation. From both ^{1}H NMR and elemental analysis, it was shown that this copolymer contained 42% units from monomer 1 of which 92% were ring opened and 58% styrene units.

To study the effect of ring size on the extent of ring opening, a seven-membered spirotriene, monomer 2 has been synthesized and polymerized. The starting material, cycloheptanecarboxaldehyde (5) was prepared in a 87% yield by use of Swern's procedure,[7] who reported successful synthesis of linear aldehydes from linear alcohols using activated DMSO in high yields. By use of the Kane's procedure[5] again, the cyclic aldehyde 5 was converted to the ketone 4 in large scale and in an overall yield of 23%. The ketone 4 was synthesized by Kane,[5] but mp, NMR, IR, and elemental analyses were not reported in the corresponding papers.

By use of the Wittig reaction[4] discussed previously, the precursor, ketone 4 was converted to the desired monomer 3-methylenespiro[5,6]dodeca-1,4-diene (2) in a 69% yield.

Homopolymerizations of 2 were carried out at different temperatures by bulk and solution methods (Table 3). In the bulk polymerization it was observed that when temperature was increased from 100 to 130°C, the extent of ring opening also increased from 69 to 100%, which was further indication that higher temperatures favored the ring opening process. Compared with monomer 1, the polymerization of the seven-membered monomer 2 gave higher extent of ring opening; for example, at 130°C monomer 2 underwent almost 100% ring opening during polymerization while monomer 1 only underwent 79% ring opening during polymerization. Apparently the relief of the higher strain energy in the seven-membered ring enhanced the ring opening process.

As predicted, in the solution polymerizations, monomer 2 gave polymers with higher extent of ring opening; for instance, at 100°C monomer 2 was polymerized in 80% (by wt) benzene to give a polymer in which 93% of the units was ring-opened, while bulk polymerization yielded a polymer with only 69% of the units ring opening.

In a copolymerization experiment, equal molar amounts of monomer 2 and styrene were copolymerized at 130°C in the presence of 2 mol% of t-butyl peroxide as the initiator for 4 h. A solid polymer obtained in a 88% conversion was isolated after purification. From elemental analysis, it was shown that this copolymer contained 56% units from monomer 2 of which 100% were ring-opened units and 44% styrene units. The higher percentage of monomer 2 units (56%) in its styrene copolymer than that of monomer 1 (42%) in its styrene copolymer indicated monomer 2 was more reactive than monomer 1 due to the higher ring strain energy.

Thus, it has been demonstrated that the spirotrienes, 1 and 2, are useful monomers that can be homo- or copolymerized with various extent of ring opening depending on the conditions, such as temperature or concentration, by a free radical mechanism to introduce the p-phenylene linkages into the backbones of addition polymers which are difficult to obtain by conventional condensation polymerizations. Furthermore, the incorporation of p-phenylene groups can be expected to produce polymers and copolymers with unusual chemical and physical properties for many applications. Although these monomers undergo shrinkage during polymerization since only one ring

Table 3. Homopolymerization of 2 at Various Temperatures

BULK POLYMERIZATION

INITIATOR (mol %)		T(°C)	TIME (h)	POLYMER YIELD(%)	SOLUBILITY	EXTENT OF RING OPENING[a](mol %)	VISCOSITY $[\eta]$ (dL/g)
Benzoyl peroxide	2	100	20	80	Soluble in $CHCl_3$, C_6H_6	69	0.631[c]
t-Butyl peroxide	2	130	4	100	Insoluble in common solvents[b]	100	

SOLUTION POLYMERIZATION

INITIATOR (mol %)		T(°C)	TIME (h)	POLYMER YIELD(%)	SOLUBILITY	EXTENT OF RING OPENING[a](mol %)	VISCOSITY $[\eta]$ (dL/g)
Benzoyl peroxide	4	100	72	50	Soluble in $CHCl_3$, C_6H_6	93	0.619[c]
t-Butyl peroxide	4	130	20	100	Insoluble in common solvents[b]	100	

[a] From 200 MHz ^{1}H NMR study
[b] Below 80°C
[c] In benzene at 30°C

is opened for every new bond that is formed, the introduction of a second ring into the monomer so that double ring opening will take place which will be expected to produce related polymers with expansion in volume on polymerization.

ACKNOWLEDGEMENTS

The authors are greatful to the Polymer Program of the National Science Foundation and the Office of Naval Research for partial support of this research.

REFERENCES

1. W. J. Bailey, Polym. J., 17(1), 85 (1985).

2. L. A. Errede, J. Polym. Sci., 49, 253 (1961).

3. B. Maillard, D. Forrest, and K. U. Ingold, J. Am. Chem. Soc., 98, 7024 (1976).

4. R. Greewald, M. Chaykovsky, and E. J. Corey, J. Org. Chem., 28, 1128 (1963).

5. V. V. Kane, Synth. Commun., 6, 237 (1976).

6. D. F. Murray, J. Org. Chem., 48, 4860 (1983).

7. K. Onura, and D. Swern, Tetrahedron, 34, 1651 (1978).

8. M. de Botton, Bull. Soc. Chim. Fr., 7-8, 1773 (1975).

9. J. W. van Straten, L. A. M. Turkenburg, W. H. de Wolf, and F. Bickel-haupt, Recl. Trav. Chim. Pays-Bas, 104, 89 (1985).

10. Auspos, Burnum, Hall, Hubbard, Kirk, Shoefgen, and Speck, J. Polym. Sci., 15, 19 (1955).

VARIETY OF ORGANOMETALLIC AND METAL-CONTAINING POLYMERS

Charles E. Carraher, Jr. and Charles U. Pittman, Jr.*

Department of Chemistry, Florida Atlantic University
Boca Raton, FL 33431, and *Department of Chemistry
University/Industrial Chemical Research Center
Mississippi State University, Mississippi State, MS 39762

INTRODUCTION

The National Research Council has noted that the major impediment to progress in many areas including computers and biomedical science is the lack of suitable materials. Both the NSF and NRC have highlighted this need for new materials whose properities meet specific specifications. One area singled out for special attention has been organometallic polymers. This review will emphasize particular aspects of such metal-containing polymers including general synthetic routes, potential and actual applications and problems.

There exists a wide variety metal-containing polymers. The metals may exist in several oxidation states and geometries. Here we will emphasize polymers that contain the metal either as an integral part of the backbone or as a part of a pendant moiety. Topics dealing with polysilicones, simple metallic salts (such as sodium polyacrylate), metalloids and metal-like systems (such as polyacetylenes) and polyphosphazenes will not be emphasized in this report.

Recent reviews of aspects of metal-containing polymers are found in references 1-6. Additionally, excellent reviews exist on polymer-bound metal complexes (7-15), preparation of vinyl organometallic polymers (16), the preparation of organometallic polymers by polycondensation (17-19), conductivity of organometallic polymers (20-23), metallocene polymers (24-26), and organotin polymers for antifouling coatings (27).

This is the first of two chapters. This chapter describes, in broad terms, the variety of metal-containing polymers thus far made and a number of factors that may limit this range. The second chapter describes a number of actual applications for organometallic polymers.

It is useful to classify organometallic polymers by the type of reaction that produces the polymer, namely condensation, coordination and addition polymerization. Superimposed on this division is the type of polymerization kinetics found, mainly stepgrowth or chain. As with organic polymers and polymerization there is a large, but not total, overlap between the terms stepgrowth kinetics and condensation polymers and with chain kinetics and addition polymers. Exceptions are abundant. The interfacial condensation of many organometallic dihalides to form condensation polymers, such as 10, occurs rapidly, following chain kinetics. Many ring opening polymerizations form condensation-type polymers by chain kinetics. We will divide the topic according to condensation, coordination, and addition polymerization but also relate the type of kinetics when appropriate. Numerous routes have been employed in the synthesis of organometallic polymers. We will focus on only some of the major routes. In general, equilibrium control is important in the formation of typical organic polymers but it is often of greater importance in the synthesis of organometallic polymers. The organometallics generally have a greater lability, thus the nature of the resulting products is often determined by competing equilibria. Many of the reactants and products can undergo rapid scrambling reactions in which parts of the same or adjacent molecules exchange giving equilibrium products.

Condensation Polymerization

Many of the natural and synthetic polymers are formed by condensation processes. A brief description of some of the more common condensation processes is given below employing examples to illustrate the particular procedure. Some examples that illustrate the breath of the use of the condensation process in effecting such syntheses include preparation of the industrially important polysiloxanes. They represent condensation polymers.

$$R_2SiCl_2 \xrightarrow[-2HCl]{H_2O} R_2Si(OH)_2 \longrightarrow \underset{\underset{R}{|}}{\overset{\overset{R}{|}}{(Si-O)}} + H_2O$$
$$\underline{1}$$

The formation of many boron-containing polymers follow the condensation process. This is illustrated below for the synthesis of borazoles (28), 2, cyclic aminoboranes, 4, and boron nitrate, 5. Boron nitride, 5, is a good example of the potential selected advantages that may occur with organometallic and inorganic polymers. It has a Young's modulus/density ratio of 20 (compared to 4 for iron and glass) and a tensile strength/density ratio of 8 (compared to 1.7 for glass and 0.3 for iron).

Boron nitride retains its high tensile strength and Young's modulus to 800 C
in an oxidizing atmosphere. It's layered structure is actually similar to
that of graphite. Another group of metal containing condensation polymers,
10, can be envisioned as direct analogues of typical condensation
polymerizations (see 9) that form polyesters and nylons. Work in this area
grew partially out of the observation that many organometallic halides and
metal halides undergo Lewis acid-base reactions such as hydrolysis, in a
manner similar to organic acid chlorides (29).

$$R - \overset{\overset{\displaystyle O}{\|}}{C} - Cl \; + \; H_2O \; \longrightarrow \; R - \overset{\overset{\displaystyle O}{\|}}{C} - OH \; + \; HCl$$

6

$$\overset{|}{\underset{|}{M}} - Cl \; + \; H_2O \; \longrightarrow \; \overset{|}{\underset{|}{M}} - OH \; + \; HCl$$

7

$$R - \overset{\overset{\displaystyle O}{\|}}{C} Cl \; + \; R - NH_2 \; \longrightarrow \; RC \overset{\overset{\displaystyle O}{\|}}{} - \overset{\overset{\displaystyle H}{|}}{N} - R'$$

8

$$Cl - \overset{\overset{\displaystyle O}{\|}}{C} - R - \overset{\overset{\displaystyle O}{\|}}{C} - Cl \; + \; H_2N - R - NH_2 \; \longrightarrow \; \left(\overset{\overset{\displaystyle O}{\|}}{C} - R - \overset{\overset{\displaystyle O}{\|}}{C} - \overset{\overset{\displaystyle H}{|}}{N} - R - \overset{\overset{\displaystyle H}{|}}{N} \right)$$

9

Condensation Polymer

$$Cl - \overset{\overset{\displaystyle R}{|}}{\underset{\underset{\displaystyle R}{|}}{M}} - Cl \; + \; H_2N - R - NH_2 \; \longrightarrow \; \left(\overset{\overset{\displaystyle R}{|}}{\underset{\underset{\displaystyle R}{|}}{M}} - \overset{\overset{\displaystyle H}{|}}{N} - R - \overset{\overset{\displaystyle N}{|}}{N} \right)$$

10

Condensation
Organometallic Polymer

Figure 1 depicts some of the structures of polymers thus far synthesized where
M can be Si, Ge, Sn, Pb, Ti, Zr, Hf, As(R_3M_2), and Sb(R_3M_2), Mn and Bi(R_3MX_2)
(see next page). Only a few of the possible difunctional monomers possible are
shown in Figure 1. Diamines, dicarboxylic acids, diols, dihydrazones, dioximes,
and dithiols are represented. One sees that practically an unlimited variety
of polymers may be envisioned. Over the years a rich variety of structures
have been prepared but outside of the well known polysiloxanes this area is
understudied. More work should be performed here.

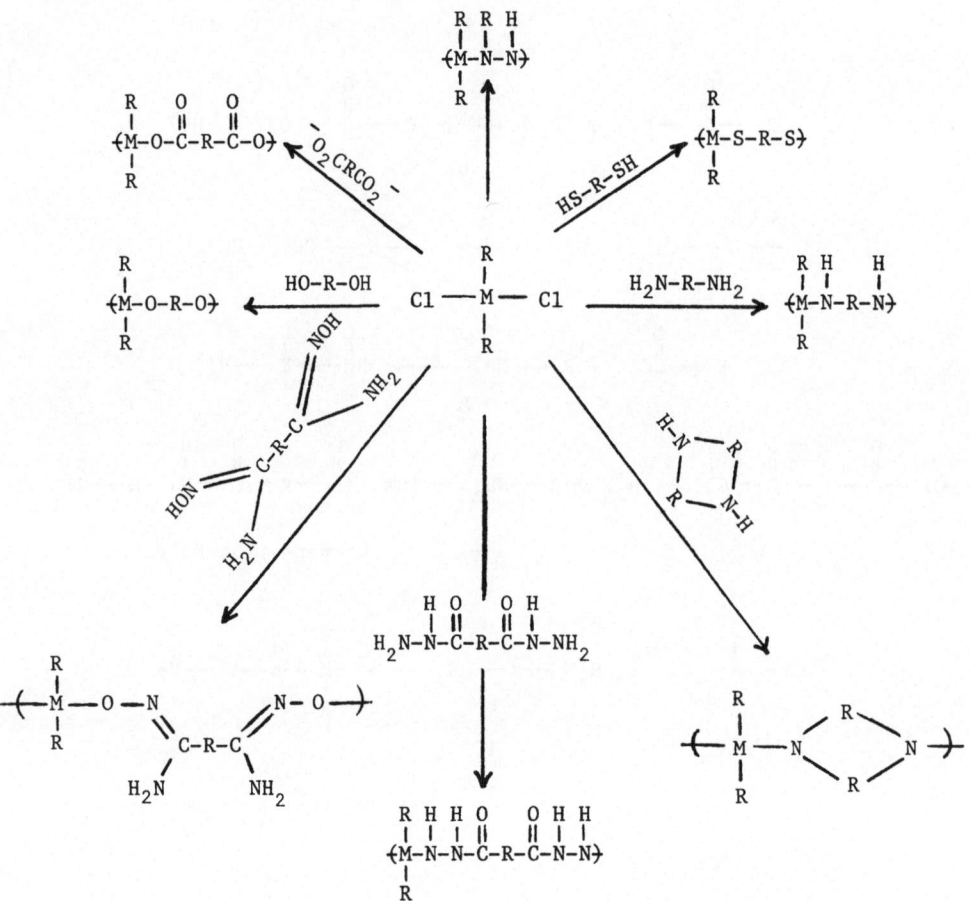

Figure 1. Representative Structures of Condensation Polymers.
M can be Si, Ge, Sn, Pb, Ti, Zr, Hf, As, Sb, MN, or Bi.

Condensation reactions have also been effected between metal halides and preformed polymers that contain Lewis acid or base sites. These preformed polymers include polyacrylonitrile, poly(vinyl alcohol), salts of carboxylic and sulfonic acids, polysaccharides, polyethyleneimine, and poly(vinyl amine). The use of metals possessing two or more reactive sites typically results in crosslinked products, 11, whereas reaction with metals containing a single reactive site yield soluble, linear products, 12, where the metal is pendant. Exceptions occur where the tendency to form internal cyclic structures is strong as the case of the reaction between xylan and dihalodiorganostannenes (30) (see 13).

The site of reaction can also occur away from the metal atom. This has particularly been exploited with Lewis acid and base metallocenes (true sandwiches) of form 14a where M = Ru, Fe, Co X , Ni.

A large number of metallocene derivatives have been subjected to polycondensation (25). Here, the Lewis acid or base could be another metal-containing reaction such as Cp_2TiCl_2 or Bu_2SnCl_2 or typical organic reactants such as diacid chlorides, doils, diamines, and disocyanates. The polyester, 14b, is a typical polycondensation product.

The formation of shishkabob macromolecules with the stacked phthalocyanine structure (15, M = Si, Ge) has been accomplished (31). These structures illustrate another general class of condensation polymers. These materials can be dissolved in strong acids, spun into fibers and doped with iodine or NO^+X^- ($X = BF_4^-$, SbF_6^-, PF_6^-) to a level where about 50% of the metals are oxidized. Conductivities as high as 1.5×10^{-1} ohm^{-1} cm^{-1} have been reported (31).

Metal-containing polymers can be derived from cyclic and cage-like monomers with polymerization occurring through the opening of a sigma bond in the cyclic or cage compound. The formation of polydimethylsiloxanes, 17, from octamethyltetrasiloxane, 16, is a well known illustration of this approach. Reaction kinetics are of the chain type and thus some authors will place each reaction under the "addition" heading. Such polymerizations reduce the translational entiopy but torsional and vibrational entropy may be increased because of the greater bond mobility in the less ligid chain.

where M = Si, Ge, Sn

Coordination Polymerization

The line between condensation, coordination, and addition is not always very clear. For instance, the reaction product between 2,6-dihydroxy-1,4-benzoquinone, 18, and a metal halide is typically classified as a coordination product, 19, yet it could also be argued that it is a condensation reaction.

The formation of coordination polymers occurs typically through three routes.

(a) Preformed metal coordination complexes (such as 20) polymerize through functional groups where the actual polymer-forming step may be a condensation, addition, or some other type of reaction.

The vinyl sandwich compounds and many other compounds considered under the general topic of addition polymerization can also be considered as coordination compounds (for instance, in 23 the iron is "coordinated" between two cyclopentadienyl rings while the polymerization of 22 is certainly an addition mechanism).

(b) A metal ion may be coordinated by preformed polymers containing chelating groups (for instance 26).

(c) Polymers form through chelation at the metal atom site (for instance 27). Typically the chelating ligand is considered to be a Lewis base that donates electron pairs. Table 1 contains a brief listing of ligands employed in the formation of coordination polymers.

Such coordination polymers have played an integral role in the life of mankind as dyes and catalysts (such as hemoglobin). Today they form the basis for many polymer-based reagents. Table 1 is given on the next page.

The drive for the synthesis and characterization of coordination polymers was originally catalyzed by work supported and conducted by the US Air Force in its search for materials which exhibited high thermal stabilities. Attempts to prepare stable, tractable coordination polymers which would simulate the exceptional thermal and/or chemical stability of model monomeric coordination compounds (such as copper ethylenediaminobisacetylacetonate(II) or copper phthalocyanine(I)) have been disappointing at best. Typically only short chains were formed and the thermal stability associated with the monomeric chelate did not seem to stand up in the corresponding coordination polymers prepared in the early work. These polymers were insoluble and high

Table 1. Chelates utilized in the synthesis of coordination polymers through chelation with metal ions

Chelate Group	Representative Structure	Chelate Group	Representative Structure
Bis-1,2-amino acids		Bis-hydroxy nitrogen heteroaromatics	
Bis-O-amino-phenols		Bis-O-hydroxy Schiff bases	
Bis-diamines	ETC.	Bis-imides	
Bis-β-diketones (Bis-1,3-diketones)		Bis-thiooxamides	
		Bis-thiopicolin-amides	
Bis-1,2-dioximes		Bis-thiosemicarba-zones	
		Bis-xanthanes	
Bis-dithiocarba-mates		Cyclophospha-zenes (with chelating groups)	
Bis-α-hydroxy acids		Dicarboxylates	
Bis-O-hydroxyazos		Dimercaptodi-ethers	
Bis-1,2-hydroxy-ketos		Phosphinous anions	
		Bis-o-nitrites and 1,2-dinitriles	

molecular weight species were not obtained since precipitation from solution occurred at an early stage of polymerization. While a hugh number of organic molecules can form coordination polymers by acting as chelating agents, only a few representative examples will be considered here.

Addition of anydrous nickel acetylacetonate to a methanolic solution of sodium tetrathiolate followed by addition of a methanolic solution of NBu_4X under argon (32) gave an amorphous precipitate. This polymer, 29, had a conductivity of only $10^{-4} S \cdot cm^{-1}$. This $\{[TTF-Ni(Bdt)](NBu_4)_x]\}_n$ polymer was then oxidized by adding a methanolic solution of bromine to the polymer suspension in methanol. The conductivity of the polymer was then elevated to $20 S \cdot cm^{-1}$. This suggests that a partial oxidation state, attributed to O_2 oxidation in the first case and Br_2 oxidation in the second, is necessary for high conductivity in these polymers. Metal poly(benzodithiolenes), like 29, have been prepared (33) from divalent ions such as Co^{2+}, Ni^{2+}, Fe^{2+}, and Cu^{2+} on reaction with the benzene-1,2,4,5-tetrathiol, 28. All of the polymers studied in this family were found to be paramagnetic conductors with conductivities ranging from 10^{-4} to $10^{-1} S \cdot cm^{-1}$. The tetrathiosquarate dianion, 30, also undergoes polychelation when reached with Ni^{2+} or Pd^{2+} salts to give materials with structure 31 (34). These Ni and Pd Polymers have degrees of polymerization of 10 and 25 and conductivities of $5 \times 10^{-1} S \cdot cm^{-1}$, respectively. The Pt-based polymer can be synthesized by reacting $K_2C_4S_4$ with K_2PtCl_4 or $(PhCN)_2PtCl_2$ to give a blackish green complex having a 1:1 ratio of $Pt:C_4S_4$ and a conductivity of $6 \times 10^{-7} S \cdot cm^{-1}$.

In general, metal coordination complexes of macrocyclic ligands, such as phthalocyanine, crystallize in stacks to afford linear chains which have the ability to conduct electricity along the stack axis. Phthalocyanine polymers with metal-metal linkages can be prepared by reacting transition metal salts with tetranitriles (34, 35), 1,2,3,4,5-tetracyanobenzene (36, 37), pyromellitic dianhydride (38, 39), or tetracarboxylic acid derivatives (38, 40). These polymers have structures generally represented by 32 where the circles represent the specific coordinating macrocycle which surrounds the metal. There is direct metal-metal bonding interaction along the stack direction. These polymers are semiconducting and exhibit pressed pellet conductivities from 10^{-1} to 10^{-1} S·cm^{-1} depending on the metal, heterocycle and state of doping.

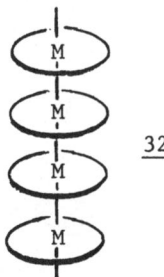

32

Phthalocyanine polymers containing bridging ligands between the metal atoms conduct electricity through their metal-ligand-metal linkages. Dehydration (a polycondensation) of phthalocyanine complexes of silicon, germanium, and tin produces face-to-face stacks of oxygen-bridge metal phthalocyanine units as shown earlier in structure 15 (41, 42). Upon oxidation with iodine, a mixed valence polymeric cation is formed with intercalated I_3^- ions between the chains. Conductivities ranged from 2×10^{-1} S·cm^{-1} for $[SiPcOI_{1.40}]_n$ and 1×10^{-1} S·cm^{-1} for $[GePcOI_2]_n$ to 2×10^{-4} S·cm^{-1} for $[SnPcOI_{5.5}]_n$. The low conductivity of the tin polymer is attributed to a larger distance between the phthalocyanine units. Electrical conductivities on the order of 10^{-2} S·cm^{-1} have been observed in undoped linear complexes with cyanide-bridging ligands, e.g., $(PcCoCN)_n$. These values compare to doped metal-phthalocyanine polymers with oxygen bridges (43).

Stacked metal chain complexes without a macrocyclic ligands are also known. Tetracyanoplatinate stacked complexes resemble the metal phthalocyanine polymeric structure, except coordinating macrocycles do not chelate each metal atom. Instead stacks of square-planar $Pt(CN)_4$ units are formed with anions between the stacks. The cyano groups surrounding Pt are staggered with respect to those coordinating the adjacent Pt atoms. These well studied materials (44-46) are one-dimentional conductors with conductivities greater than 1 S·cm^{-1}. For example, the electrolysis of a solution of Rb_2SO_4 and $Rb_2Pt(CN)_4$ gives reddish-brown crystals of $Rb_2Pt(CN)_4H_2O$ at the electrodes.

Addition Polymerization

Addition polymerization processes can be subdivided into several types. The first type involves a decrease in the number of pi bonds and a corresponding increase in the number of sigma bonds. The total valency of each atom remains constant. The formation of polymeric selenium dioxide is of this type (eg. 33).

33

More typical examples involve the polymerization and copolymerization of metal-containing vinyl monomers such as pictured in Table 2 (next page). These reactions have been extensively studied and have been recently reviewed (for instance 16, 47-49). The first organometallic monomer which contained a

Table 1. Vinyl and Alkynyl Organometallic Monomers Which Have Been Prepared

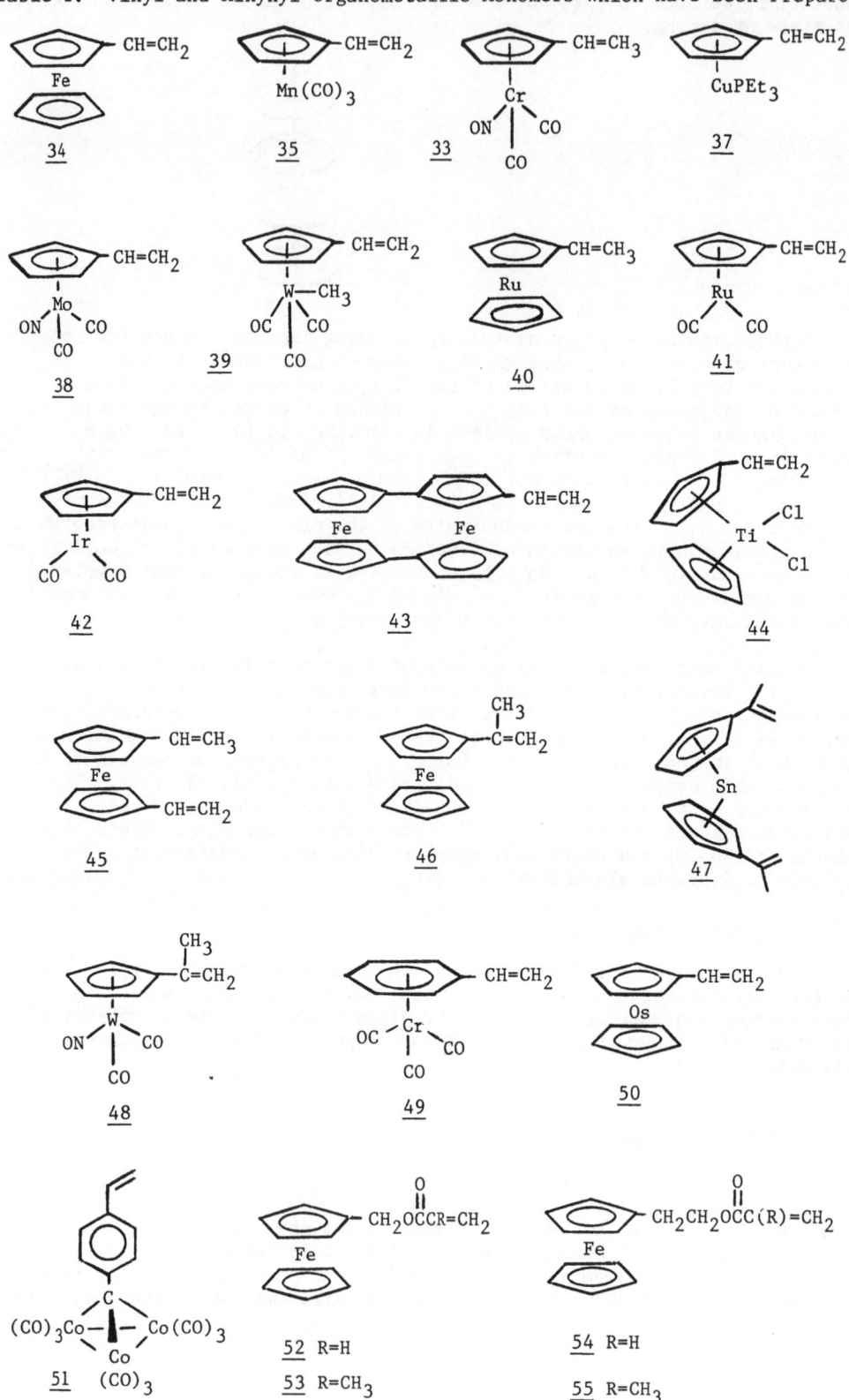

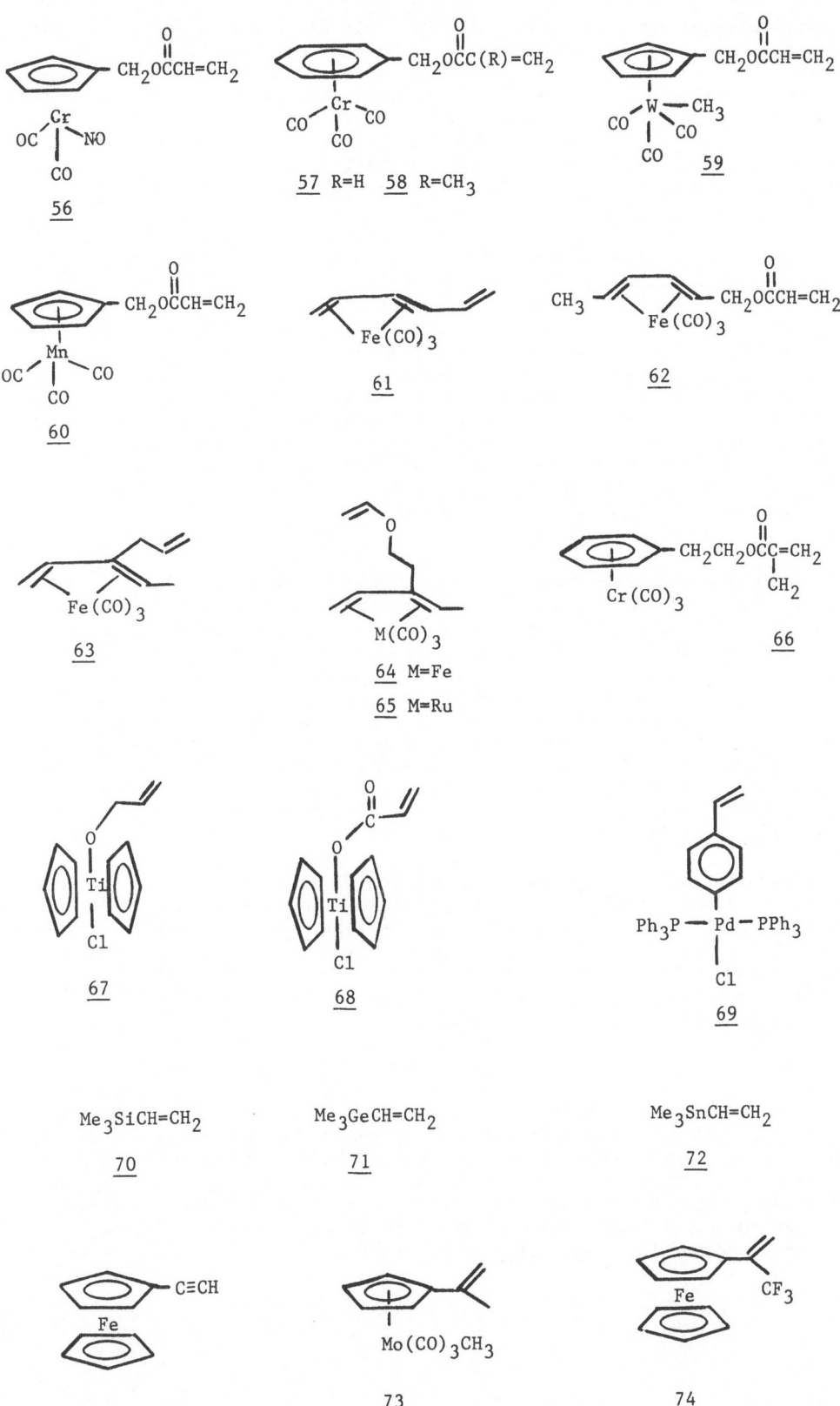

56

57 R=H 58 R=CH$_3$

59

60

61

62

63

64 M=Fe
65 M=Ru

66

67

68

69

Me$_3$SiCH=CH$_2$

70

Me$_3$GeCH=CH$_2$

71

Me$_3$SnCH=CH$_2$

72

73

74

105

transition metal was vinylferrocene, 34. This monomer was prepared in 1955
(50) and its polymerization behavior has been extensively studied under
radical (50-53), cationic (54) and Ziegler-Natta conditions (55). Monomer 34
is inert to anionic initiation (56). Several other vinly-type organometallic
monomers, including vinylcymantrene, 35, (57-58) and vinylcynichrodene, 36,
(60-61) have more recently been prepared. They have been homopolymerized and
copolymerized with a variety of electron-donating and electron-attracting
organic comonomers (i.e., styrene, methyl mathacrylate, acrylonitrile and N-
vinylpyrrolidine) (58-61).

The effect that organometallic functions exert in viniyl polymerizations
is beginning to become clear in some instances. A transition metal with its
various readily available oxidation states and large steric bulk might be
expected to exert unusual electronic and steric effects during polymerization.
Here, we will consider the vinylcyclopentadienyl monomers using
vinylferrocene, 34, as an introductory example. Its homopolymerization has
been initiated by radical (50, 51, 62-65) cationic (54) coordination (66) and
Ziegler-Natta (54) initiators. Unlike the classic organic monomer, styrene,
vinylferrocene undergoes oxidation at iron when peroxide initiators are
employed. Thus, azo initiators (such as AIBN) are commonly used. The
stability of the ferricinium ion makes ferrocene readily oxidizable by
peroxides whereas styrene, for example, would undergo polymerization. This is
true of many other organometallic systems.

Unlike most vinyl monomers, the molecular weight of poly(vinylferrocene)
does not increase with a decrease in initiator concentration (51). This is
the result of vinylferrocene's anomalously high chain-transfer consant
(C_m = 8 x 10^{-3} versus 6 x 10^{-5} for styrene at 60°C). Finally, the rate law for
vinylferrocene homopolymerization (e.g. r_p = k[VF]^1[In]^1) is the first order
in initiator in benzene solvent (52). Thus, intramolecular termination
occurs. Mössbauer studies supported a mechanism involving electron transfer
from iron to the growing chain radical center giving a zwitterion which
terminates the chain. Subsequent decomposition results in a high spin Fe(III)
complex (53). In dioxane the rate law was "normal" (e.g. first order in
monomer, half order in initiator) indicating a bimolecular termination
mechanism.

The electron rich cyclopentadienyl ring in vinyl monomers 34-44 (Table 2)
is able to stabilize adjacent positive charge. This has long been noted in
ferrocene systems where the exceptional stability of α-ferocenyl carbenium
ions is well known. Thus, the vinyl group in these monomers might be expected
to be quite electron rich and behave as such, both in radical and cationic
initiated homo- and copolymerizations. This is illustrated by the ready
cationic polymerization of 1,1'-divinylferrocene, 45. Molecular weights up to
35,000 have been obtained using BF$_3$(Et$_2$O) initiation (67). This
polymerization forms cyclolinear structures 75, due to intramolecular
electrophilic attack of the α-ferrocenylcarbenium ion on the adjacent vinyl
group (67, 68). Cationic initiation of some of the isopropenyl organometallic
monomers occurs with some difficulty as discussed by Rausch and Pittman (69-
71). These (α-isopropenylcyclopentadienyl)metal monomers can be considered as
analogs of α-methylstyrene. However, despite this analogy 47, 73, and 74 were
not readily homo or copolymerized under a variety of cationic conditions
and high molecular weight homo- or copolymers have not been achieved.
Likewise, radical initiation of 61 does not readily lead to high polymers due

to the inability of <u>76</u> to propagate successfully. As might be expected, these electron-rich vinyl monomers resist anionic initiation.

<u>75</u>

<u>76</u>

The preparation of the extensive series of vinyl monomers listed in Table 1 involves a knowledge of organometallic chemistry. Vinylcymantrene, <u>35</u>, as well as monomers <u>34</u>, <u>36</u>, <u>40</u>, <u>45</u>, and <u>46</u> can be prepared by Friedel Crafts acylation of η^5-cyclopentadienyltricarbonylmanganese followed by $NaBH_4$ reduction of the acetyl derivative to the alcohol and dehydration (58, 59). This classic organic route, however, cannot be employed successfully with many organometallic species. The organometallic moieties can be destroyed under Friedel Crafts conditions. Thus, other methods were needed. Formylcyclopentadienide was made and reacted with $W(CO)_6$ followed by treatment of the resulting tungsten-centered anion with methyl iodide (72). The resulting aldehyde was converted to the vinyl monomer via the Wittig reaction. Since Wittig chemistry is not always easy to carry out, or scale up, a more general route involving vinylcyclopentadienyl lithium has been used (71, 73, 74). Starting with 6-methylfulvene one generates vinylcyclopentadienyl lithium which, in turn, can react with various organometallics to give vinyl monomers such as <u>37</u> or <u>42</u> (similarly, 6,6-dimethylfulvene becomes to the precursor of <u>47</u> and <u>73</u>).

<u>35</u>

$$\text{Na}^+ + \text{H-COEt (O)} \longrightarrow \text{CHO, Na}^+ \xrightarrow{\text{W(CO)}_6} \text{CHO, W(CO)}_3$$

$$\xrightarrow{\text{CH}_3\text{I}} \text{CHO, W(CO)}_3\text{CH}_3 \xrightarrow{\text{Ph}_3\text{P=CH}_2} \text{CH=CH}_2, \text{W(CO)}_3\text{CH}_3 + \text{POPh}_3$$

39

$$\text{C} \begin{smallmatrix} \text{CH}_3 \\ \text{H} \end{smallmatrix} \xrightarrow{\text{LiN(CHMe}_2)} \text{CH}_2, \text{C, H, Li}^+ \xrightarrow{(\text{ClCuPEt}_3)_4} \text{CH=CH}_2, \text{CuPEt}_3$$

37

$$\downarrow \text{Ir(CO}_3)\text{Cl}$$

$$\text{CH=CH}_2, \text{Ir(CO)}_2$$

42

Most recently, (75) vinylcymantrene, 35, was converted to $\{[\eta^5\text{-C}_5\text{H}_4(\text{CH=CH}_2)]\text{Mn(NO)(CO)}_2\}^+ (\text{PF}_6^{-1})$ using NOPF_6 followed by conversion to the novel organometallic monomer $[\eta^5\text{-C}_5\text{H}_4(\text{CH=CH}_2)]\text{Mn(NO)(S}_2\text{C}_6\text{H}_3\text{CH}_3)$ upon reaction with $(\text{HS})_2\text{C}_6\text{H}_4\text{CH}_3$. Thus, one organometallic monomer served as the precursor to another monomer having a very different structure about the metal.

Another type of addition polymerization process involves ring opening. Polymers can be derived from cyclic and cage-like ring opening polymerizations. Depending on the particular product obtained such polymers can be considered as condensation or addition polymers. Here we will consider products that generally contain the same element in the chain as addition polymers. An example is polysulfur, 78, whereas polymers such as polydimethylsiloxanes, 17, derived from octamethyltetrasiloxane, 16, have been considered as condensation polymers.

$$77 \longrightarrow (\text{S})_n$$

78

(c) Polymers can be formed by coordination to a metal where no bonds are broken but additional ones formed. This results in an increased coordination number. Thus, palladium chloride is monomeric in the vapor state but polymerized through addition kinetics in the solid state, 79. Such reactions can equally be considered as coordination reactions.

$$\text{PdCl}_2 \longrightarrow \text{Pd}\begin{smallmatrix} \text{Cl} \\ \text{Cl} \end{smallmatrix}\text{Pd}\begin{smallmatrix} \text{Cl} \\ \text{Cl} \end{smallmatrix}$$

79

The reaction of carbenes typically occur through chain-like kinetics. Thus phenylboron dichloride (80) when heated under reflux in benzene, in the presence of sodium or lithium in an inert atmosphere, gives short chain as noted below. The resulting polymer, 81, presumeably formed via a "borene" analog to carbene.

$$Ph-BCl_2 \ + \ Na \ (or \ Li) \ \longrightarrow \ [Ph-B:] \ \longrightarrow \ \underset{\underset{\underset{81}{Ph}}{|}}{\left(B \right)_n}$$

PROBLEMS

The following is a very brief description of some of the problems encountered in the generation of metal containing polymers.

1. Compared to carbon chemistry, possible mechanistic pathways are more numerous with most metals to include all varieties of associative, dissociative, interchange and exchange pathways.

2. Many of the metal-containing reactants and products are highly reactive with water and air, undergoing disproporation interchange and exchange reactions and readily undergoing rearrangement reactions.

3. The synthesis of many desired monomers is difficult or not yet achieved.

4. Properties of many possible monomers are unknown, little known, or misknown (i.e. wrong), thus the researcher must become familiar with at least some of the basic properties of the employed metal-containing reactants including proper comonomers, realistic reaction conditions, analysis procedures, and proper (and safe) handling procedures. Much of this familiarity must come from the first hand experience of the investigator in addition to the literature.

5. Because of the highly reactive nature of many of the metal-containing reactants, selection of reaction conditions are critical and often seriously limited.

6. Most of the metal-containing polymers available at this time are difficult to process and many exhibit very limited softening and solubility ranges Many are highly crystalline and the site about the metal-atom itself is sometimes rigid.

7. Degradation, chemical, physical, and thermal, generally occurs through a series of routes and is complex in comparison with the typical organic polymers.

8. Last, applications will be limited with regard to the cost and availability of both the metal and metal-containing reactant. Polymers containing relatively abundant metals such as tin, lead, silicon, iron, and cooper may be used in large scale applications but polymers containing scarce metals such as platinum, rhodium, and hafnium will be delegated to small scale, but potentially quite important, applications. The demand for specialty materials is expanding at a rate corresponding to the growth

in the complexity of our technology. Thus, there is a large real and potential market for metal-containing polymers.

REFERENCES

1. J. E. Sheats, C. E. Carraher, Jr., and C. U. Pittman, Jr., (eds.), "Metal-Containing Polymeric Systems", Plenum Press, N.Y. pp. 1-523 (1985).
2. C. E. Carraher, Jr., J. E. Sheats, and C. U. Pittman, Jr., (eds.) "Advances in Organometallic and Inorganic Polymer Science", Marcel Dekker, Inc., N.Y. and Basel, pp. 1-449 (1982).
3. C. E. Carraher, Jr., J. E. Sheats, and C. U. Pittman, Jr., "Organometallic Polymers", Academic Press, N.Y., pp. 1-346, (1978).
4. J. E. Sheats, C. U. Pittman, Jr., and C. E. Carraher, Jr., Chemistry in Britain, 20(8):709-714 (1984).
5. J. E. Sheats, J. Macromol. Sci. A16(6):1173, 1980.
6. C. U. Pittman, Jr., Chemtech, 1:416-423 (1971); K. A. Andrianov, Metallorganic Polymers, Wiley, N.Y. (1965).
7. C. U. Pittman, Jr. in: "Comprehensive Organometallic Chemistry," G. Wilkinson, F. G. A. Stone, and E. W. Abel (eds.), Pergamon Press, Oxford, Vol. 8, Chapter 55, 553-608 (1982).
8. F. R. Hartley, Supported Metal Complexes, D. Reidel Publishers, pp. 1-318 (1985).
9. Y. Chauvin, D. Commereuc, and F. Dawans, Prog. Polymer Sci., 5:95 (1977).
10. C. U. Pittman, Jr. and G. O. Evans, Chemtech, 3:560 (1973).
11. C. U. Pittman, Jr. in: P. Hodge and D. C. Sherrington (eds.) "Polymer-Supported Reactions in Organic Synthesis", Wiley, Chichester, pp. 249-291 (1980).
12. N. Toshima, Yuki Gosei Kagaku Kyokaishi, 36:909 (1978); Chem. Abst., 90:104033 (1979).
13. V. M. Akhmedov, A. A. Medzhidob, and A. G. Azizov, Asserb. Khim. Zh., 122 (1979), Chem. Abst., 92:11692 (1980).
14. A. L. Robinson, Science, 194:1261 (1976).
15. W. T. Ford (ed.), "Polymeric Reagents and Catalysts", Amer. Chem. Soc., Washington, DC, pp. 1-295 (1986).
16. C. U. Pittman, Jr., in: "Organometallic Reactions", E. I. Becker and M. Tsutsui (eds.), 6:1-62, Marcel Dekker, NY (1977).
17. C. E. Carraher, Jr., Chemtech, 744 (1972).
18. C. E. Carraher, Jr., J. Chem. Ed., 58(11):921 (1981).
19. C. E. Carraher, Jr., et al., in: "Interfacial Synthesis", C. E. Carraher, Jr. and J. Preston (eds.), Vol. III, Plenum, NY (1982).
20. B. A. Bolto and J. Katon, (ed.), "Organic Semiconducting Polymers", Marcel Dekker, pp. 199-257 (1968).
21. C. Pecil, G. Zerbi, R. Bozio, and A. Girlando, (eds.) Proceedings of the International Conference on the Physics and Chemistry of Low-Dimentional Synthetic Metals (ICSM 84), Abano Terme, Italy, June 17-22 (1984).
22. A. J. Epstein and E. M. Conwell (eds.), Proc. of the Internat. Conf. on Low Dim. Conductors, in: "Mol. Cryst. Liq. Cryst.", 82:881-1028 (1982).
23. J. S. Miller (ed.), "Extended Linear Chain Compounds", Vol. 1-3, Plenum Press, (1982).
24. E. W. Neuse, Encycl. Polym. Sci. Tech., 8:667 (1968).

25. E. W. Neuse and H. Rosenberg, <u>Metallocene Polymers</u>, Marcel Dekker, NY (1970).
26. E. W. Neuse, in ref. 2, pp. 3-72.
27. R. V. Subramanian and K. N. Samosekharan, in ref. 16, pp. 73-93.
28. B. J. Aylett, "Inorganic Polymers", B. Currell (ed.), Royal Society of Chemistry, London, 1984, pp. 35-45.
29. C. Carraher, <u>J. Chem. Ed.</u>, 58(11):921-934 (1981).
30. Y. Naoshimon and C. Carraher, Unreported results.
31. C. W. Dirk, T. Inabe, K. F. Schoch, and T. J. Marks, <u>J. Amer. Chem. Soc.</u>, 105:1539 (1983).
32. J. Ribas, P. Cassoux, and F. Gallais, <u>C. R. Acad. Sci.</u>, Paris, 293 (1981).
33. C. W. Dirk, M. Bousseau, P. Barrett, F. Moraes, F. Wudl, and A. Heeger, <u>Macromolecules</u>, 19:266 (1986).
34. F. Gotzfried, W. Beck, A. Lerf, and A. Sebald, <u>Agnew Chem. Int. Ed. Engl.</u>, 18:463 (1979); G. Kossmehl and M. Rohde, <u>Macromol. Chem.</u>, 178:715 (1977).
35. W. D. Bascom, R. L. Cottington, and T. Y. Ting, <u>J. Mater. Sci.</u>, 15:2097 (1980).
36. C. J. Norrel, et al., <u>J. Polym. Sci., Polym. Chem. Ed.</u>, 12:913 (1974).
37. W. Hanke, <u>Z. Chem.</u>, 6:69 (1966).
38. C. S. Marvel and J. H. Rassweiler, <u>J. Am. Chem. Soc.</u>, 80:1197 (1958).
39. L. Kreja and A. Plewka, <u>Electrochim. Acata</u>, 25:1283 (1980).
40. A. S. Akopor, T. N. Lomova, and B. D. Berezin, <u>Izv. Vyssh. Uchebn. Zaved., Khim. Khim. Teknol.</u>, 19:1177 (1976).
41. J. L. Peterson, C. S. Schramm, D. R. Stojakovic, B. M. Hoffman, and T. J. Marks, <u>J. Am. Chem. Soc.</u>, 99:286 (1977).
42. K. F. Schock, Jr., B. R. Kundalkar, and J. T. Marks, <u>Am. Chem. Soc., Org. Coat. Plast. Chem., Pap.</u>, 41:127 (1979).
43. J. Metz and M. Hanack, <u>J. Am. Chem. Soc.</u>, 105:828 (1983).
44. K. Krogmann and H. D. Hausen, <u>Z. Anorg. Allg. Chem.</u>, 358:67 (1968).
45. K. Krogmann, <u>Angew Chem. Int. Ed. Engl.</u>, 8:35 (1969).
46. L. S. Miller and A. J. Epstein (eds.), "Synthesis and Properties of Low Dimensional Materials", Ann. NY Acad. Sci., 313 (1978).
47. C. U. Pittman, Jr., <u>J. Paint Technol.</u>, 43 (1971) and <u>Chemtech.</u>, 1:416 (1971).
48. C. U. Pittman, Jr., P. Grube, and R. M. Hanes, <u>J. Paint. Technol.</u>, 46:597 (1974).
49. C. U. Pittman, Jr., in ref. 3, p. 1-11 (1978).
50. F. S. Arimoto and A. C. Haven, Jr., <u>J. Am. Chem. Soc.</u>, 77:6295 (1955).
51. Y. Sasaki, L. L. Walker, E. J. Hurst, and C. U. Pittman, Jr., <u>J. Polym. Sci. Chem. Ed.</u>, 11:1213 (1973).
52. M. H. George and G. F. Hayes, <u>J. Polym. Sci. Chem. Ed.</u>, 13:1049 (1975).
53. M. H. George and G. F. Hayes, <u>J. Polym. Sci. Chem. Ed.</u>, 14:475 (1976).
54. C. Aso, T. Kunitake, and T. Nakashima, <u>Macromol. Chem.</u>, 124:232 (1969).
55. C. R. Simionescu, <u>Macromol. Chem.</u>, 163:59 (1973).
56. C, U. Pittman, Jr. and C.C.Lin, <u>J. Polym. Sci. Polym. Chem. Ed.</u>, 17:271 (1979).
57. A. N. Nesmeyanov. K. N. Anisimov, N. E. Kolobova, and I. B. Zlotina, <u>Dokl. Akad. Nauk. SSSR</u>, 154:391 (1964).
58. C. U. Pittman, Jr. and T. D. Rounsefell, <u>Macromolecules</u>, 9:937 (1976).
59. C. U. Pittman, Jr., C. C. Lin, and R. D. Rounsefell, <u>Macromolecules</u>, 11:1022 (1978).
60. E.A. Mintz, M. D. Rausch, B. H. Edwards, J. E. Sheats, T. D. Rounsefell, and C. U. Pittman, Jr., <u>J. Organometal. Chem.</u>, 137:199 (1977).

61. C. U. Pittman, Jr., T. D. Rounsefell, E. A. Lewis, J. E. Sheats, B. H. Edwards, M. D. Rausch, and E. A. Mintz, Macromolecules, 11:560 (1978).
62. J. C. Lai, T. Rounsefell, and C. U. Pittman, Jr., J. Polym. Sci. A-1, 9:651 (1971).
63. Y. H. Chem, M. Fernandez-Refojo, and H.G. Cassidy, J. Poly Sci., 40:433 (1969).
64. C. U. Pittman, Jr., R. L. Voges, and J. Edler, Polym. Lett., 9:191 (1971).
65. C. U. Pittman, Jr. and P. L. Grube, J. Polym. Sci. A-1, 9:3175 (1971).
66. C. U. Pittman, Jr., Polym. Lett., 6:19 (1968).
67. S. L. Sosin, L. V. Jashi, B. A. Antipova, and V. V. Korshak, Vysokomol. Soedin., XII (9):699 (1970).
68. J. C. Lai, T. D. Rounsefell, and C. U. Pittman, Jr., J. Polym. Sci. A-1, 9:651 (1971).
70. C. U. Pittman, Jr. and M. D. Rausch, Pure and Appl. Chem., 58(4):617 (1986).
71. M. D. Rausch, D. W. Macomber, F. G. Fang, C. U. Pittman, Jr., T. V. Jayaraman, and R. D. Priester, Jr., in: "New Monomers and Polymers", B. M. Culbertson and C. U. Pittman, Jr. (eds.), Plenum Press, New York, 243-267 (1984).
72. C. U. Pittman, Jr., T. V. Jayaraman, R. D. Priester, Jr., S. Spencer, M. D. Rausch, and D. Macomber, Macromolecules, 14:237 (1981).
73. D. W. Macomber, W. P. Hart, M. D. Rausch, R. D. Priester, and C. U. Pittman, Jr., J. Amer. Chem. Soc., 104:884, 1982.
74. M. D. Rausch, D. W. Macomber, K. Gonsalves, F. C. Fang, Z. R. Lin, and C. U. Pittman, Jr., Polymer Materials: Science and Engineering Preprints (ACS), 49:358 (1983).
75. E. G. Miller, M. J. Naughton, and R. Heintz, Abstracts 193rd ACS National Meeting, Denver, Colorado, April 5-10, 1987, INOR 219.

APPLICATIONS OF ORGANOMETALLIC POLYMERS

Charles U. Pittman, Jr. and Charles E. Carraher, Jr.*

Department of Chemistry, University/Industry Chemical Research
Center, Mississippi State University, Ms. State, Ms. 39762 and
*Department of Chemistry, Florida Atlantic University, Boca Raton
Florida, 33431

INTRODUCTION

The number of potential applications in which the use of metal-containing polymers can be considered is, to a first approximation, limited only by one's imagination. Areas of application currently under investigation include a) catalysis, b) thermally stable materials applications, c) biological (antifungal, insecticides), d) biomedical (antibacterial, controlled release, antiviral, antitumoral), 3) additives (coatings, paper, plastics), f) electrical (conductors, semiconductors), g) photoactive materials (xeroxing type of applications), h) analytical applications, i) flame retardants, j) nonlinear optical devices, and k) preparation of ceramics. Following is a brief survey of some of the above areas including some specific illustrations.

EXAMPLES AND APPLICATIONS

Electrical Applications

The design and synthesis of conducting and semiconducting polymers is a major research focus currently. For example, the number of publications on poly(acetylene) exceeds 300 in the past year. Conductivity in metal-containing polymers is also a topic of interest. Most conducting polymers are doped systems. Doping renders most polymers air and moisture sensitive which has been a disadvantage in the rapid application of conducting polymers. Many organometallic polymers which have conducting properties have poor physical properties and are not easily characterized. An example of a "brick dust"-like conducting organometallic polymer is the complex, 1, of poly(vinylbisfulvalenediiron) with tetracyanoquinodimethan (1). Thus, despite its encouraging conducting behavior it is not useful in practice. Most of the organometallic polymers shown in Figure 1 of the preceeding chapter are semiconductors with bulk conductivities ranging from 10^{-2} to 10^{-13} $S \cdot cm^{-1}$ for both AC and DC measurements (2). They exhibit high dielectric constants and dissipation factors. However, encouraging advances are being made.

113

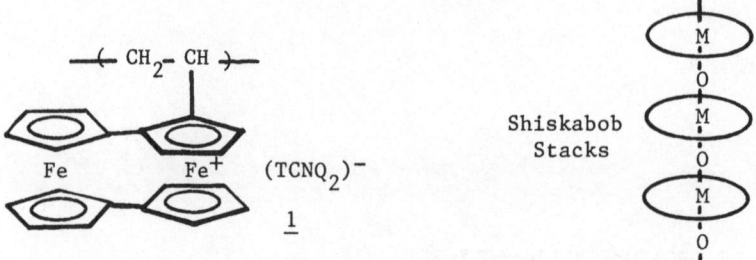

Cofacial polymers constructed from metal phthalocyanines of the type $[M(Pc)O]_n$, M = Si, Ge, Sn; Pc - phthalocyaninato, (see above) constitute a robust structurally regular class of polymers. These polymers have a "shishkabob" structure (2). They can be doped, either chemically or electrochemically, to access a wide variety of crystal and electronic structures. The electrical, optical, and magnetic properties respond to changes in the ring-ring interplanar spacing, the nature of the off-axis counterions, and the band filling (3). The linear chains conduct electricity along the stack axis (3, 4). For example, conductivities ranged from 2×20^{-2} S·cm^{-1} for $[Si(Pc)OI_{1.40}]$ and 1×10^{-1} S·cm^{-1} for $[Ge(Pc)OI_2]_n$ to 2×10^{-4} S·cm^{-1} for $[Sn(Pc)OI_{5.5}]_n$ The tin polymer has the larger interphthalocynine distance.

A phthalocyanine polymer containing pyrazine as a bridging ligand instead of oxygen, (i.e. $[Fe(Pc)(\mu\text{-pyz})]_n$) is a dark violet solid prepared from ferrous phthalocyanine and pyrazine (4). The I_2 doped polymer, $\{Fe(Pc)(pyz)]I_y\}_n$, can be prepared by either treating the suspended, undoped polymer with iodine in benzene or by precipitating the polymer from a solution of Fe(Pc)(pyz) containing iodine. The room temperature conductivity of the undoped polymer was found to be on the order of 10^{-6}-10^{-7}S·cm^{-1}(pressed pellet). The I_2 doped polymer showed an elevated conductivity of 10^{-1}-10^{-2}S· cm^{-1}(pressed pellet).

Mixed valence transition metal complexes of the dianion of oxalic acid act as semiconductors. The Fe(II) oxalates are linear chain complexes as shown in structure 2. Oxidation of $Fe(C_2O_2)(H_2O)_2$ with Br_2 gives $Fe(C_2O_2)(H_2O)_{1.4}Br_{0.6}$, a semiconducting polymer with a room temperature conductivity of 10^{-4} S·cm^{-1}. The reaction of $Fe(C_2H_2O_4)(H_2O)_2$ with I_2 results in the formation of a complex with a conductivity six orders of magnitude greater than that of the unoxidized polymer (5). The structurally related poly(metal tetrathiooxalates) have been prepared (6) using 1:1 stoichiometry of tetraethylammonium tetrathiooxalate and metal^{2+} salts (see 3). Conductivities of these complexes range from 5 to 20 S·cm^{-1} for $[NiC_2S_4]_n$ to 1 S·cm^{-1} for $[CuC_2S_4]_n$ and $[PdC_2S_4]$. These complexes are actually n-type semiconductors as evidenced by their low negative thermoelectric power coefficients of ca. -10 VK . Vapor phase I_2 oxidation of these complexes leads to a decrease in their electrical conductivity as the concentration of mobile electrons is depleted.

2

$$x[Et_4N]_2C_2S_4 + xNi[NO_3]_2 \longrightarrow \left[\cdots \right]_x + 2x[Et_4N][NO_3]$$

3

The reaction of butadiyne with cuprous chloride in dimethylacetamide under argon (7) gives the three dimensional poly(metal-yne), 4. A similar, possible linear, polymer, 5, is obtained by reacting butadiyne with cupric chloride in ammonia methanol solutions. Conductivities were found to be 10^{-9} S·cm^{-1} for 4 and 5. Upon iodine doping the conductivities are enhanced by about 10 orders of magnitude to 6.3 S·cm^{-1} for 4 and 10.2 S·cm^{-1} for 5. Both polymers 4 and 5 display metallic behavior.

4

$$\left[C \equiv C - C \equiv C - Cu \right]_n$$

5

A large number of applications appear to be developing in the area of functionalized electrodes for electrocatalysis, photoelectrocatalysis, photovoltaic cells, and specialized electrode and sensor development (8-14). For example, a [1.1]-ferrocenophane-containing polystyrene when applied as a film to the surface of a p-type semiconductor gives a photoelectrolysis device from which H$_2$ can be generated from acid at 430 mV more positive than for H$_2$ evolution from platinum surfaces under the same conditions (14).

Poly[tris(5,5'-bis[3-acrylyl-1-propoxy)carbonyl]-2,2'-bipyridine)ruthenium] is one of a series of polymerized electrochromic compounds which have recently been reported to go through a range of seven different colors from orange to cherry red. This series of polymeric complexes may be useful in generating multicolor displays in electronic display panels (15). When deposited as thin films on tin oxide supports, they have been cycled more than 10^6 times through all of possible ruthenium oxidation states (+2 → -4) with 85-90% of material remaining intact and a response time of a few tenths of a millisecond.

Analytical Reagents and Metal Ion Concentrating Agents

Formation of organometallic polymers can be utilized as a means to concentrate, isolate, purify, and separate metal-containing moieties. The natural water soluble form of uranium is the uranyl ion. It has been effectively chelated employing difunctional monomers such as terephthalic acid, polymers such as sodium polyacrylate and a wide variety of carboxylate and sulfonate resins (16, 17). Preformed poly(thiosemicarbazide), 6, is extremely efficient in complexing Cu^{2+} ions from the waste effluents from bass mills thereby forming the polymeric copper chelate 7 (18). The Cu^{2+} ions were specifically complexed with high selectivity in the presence of high concentrations of other ions. The copper is tenaciously bound in polymer 7 and is not eluted on treatment with mineral acids, but elution with 1,4-benzoquinone releases copper in purities of 96-99% in field trials. It is interesting to note that the copper-containing polymer, 7, can be used as a reagent for the oxidation of aldehydes to carboxylic acids and the coupling of nitroso compounds to azoxy derivatives.

Permanent Polydyes

A number of metal-containing polydyes have been synthesized. Thus a number of dyes, such as phenolphthalien and mercurochrome, 8, possess two basic functional groups which can be condensed with organometallic halides (19, 20). Those containing tin show a wide range of biological activity and a number of the dyes are employed in biology and medicine as cell stains. The metal polydyes, such as 9 and 10, are fluorescent and readily impregnate paints, paper, cloth, and plastics yielding fluorescent materials. Suggested uses include trace additives for identification, laser applications, permanent coloring agents in paints, paper, cloth, and plastics and in biomedical applications such as specialized stains and toxins. Such dyes have been added to coatings, fabrics, caulk, sealants, plastics, and paper materials (20).

UV Stabilizers

Compounds containing Cp_2M units where M=Fe, Hf, Zr, or Ti are stabilizers to UV radiation. Thus, inclusion of small amounts into polymer chains which in turn, are incorporated in a paint might impart to the outdoor paint better weatherability regarding light stability (21, 22).

Biological Activity

A wide variety of metal-containing polymers have been made for their potential biological activities. Only a few selected examples will be mentioned. Almost all compounds containing tin (including those in references (23-28)) are antifungal and antibacterial. Tin-containing polymers can impart mildew and rot resistance to clothing, rugs, etc. Tributyltin methacrylates have found widespread use in antifouling paints which slowly release or ablate tributyltin hydroxide (29). Tin species are toxic to sliming organisms in the biological progression of organisms involved in ship bottom fouling (30).

Organometallic polymers are being investigated as medicinal agents (31-36). Coordination of K_2PtCl_4 by poly[bis(methylamine)phosphazene] gives the platinum phosphazine polymer, 11, which shows tumor inhibitory activity against mouse P388 lymphocytic leukemia and in the Ehrlich Ascites tumor regression test. The cis-dichloroplatinum moiety has been included in the polymer repeating units by the reaction of K_2PtCl_4 with such difunctional nitrogen compounds as 1,6-hexamethylenediamine, pyrimidines, purines, and hydrazines (see structure 12 for example) (31, 35-36). At 10 to 20 µg/ml they sometimes repress replication of poliovirus I and L RNA virus but have no activity towards L929 mouse, HeLa or WISH cells. At concentrations above 30 µg/ml tumoral cell growth is stopped.

11   12

Thickening Agents

Polymeric metal phophinates having single, double or triple-bridged structures, 13-15, have been made using Al, Be, Co, Cr, Ni, Ti, and Zn (33). They form films with thermal stabilities to 450°C and the chromium(III) polyphophinates have been used as thickening agents for silicone greases to improve their high pressure physical properties.

13   14   15

Energy Transfer

Attempts have begun to use polymer matrices containing europeium chelates as the active species in the electronic energy transfer processes of lasers (38, 39). Europeium chelates are of interest because the pump energy is absorbed by the organic chelate molecule and efficiently transfered to the europeium ion. One example monomer, 16, was copolymerized with methyl methacrylate, and the resulting translucent polymer showed superfluorescence when excited by pulses from xenon flashlamps (39).

Several organometallic polymers have been used as pre-heat shields for targets in inertial-confinement nuclear fusion (40). Temperatures of 50,000,000 degrees are needed for fusion. Small hollow spheres containing deuterium and tritium are placed at the focal point of high intensity laser beams and subjected to eight trillion watts cm^2. The release of suprathermal electrons from the target's shell causes efficiency loss in fusion because the electrons heat the core prematurely. This is remedied by having a concentric shell of low average atomic number material containing 1-4 atom percent of atomic number 50-85 uniformly distributed on the molecular level. Hence, among others, polymers and copolymers of vinylruthenocene and vinylosmocene and $Ph_3PbC_6H_4CH=CH_2$ have been employed to make the concentric shell having the evenly distributed atoms in the atomic number range of 50-85.

Anomolous Fibers

A number of the polymers containing Cp_2Ti groups exhibit "anomolous fiber formation" reminiscent of metal whiskers. At low temperature their thermal stability is poor, but after initial weight loss below 250°C they retain 80% of their weight to 1200°, and significant organic portions remain intact at above 600°(41).

Composites

Several ferrocene-containing polymers have been used in the preparation of composites. For instance, ferrocene, copolymerized with terephthaldehyde and 1,1'-ferrocenedicarboxaldehyde gave thermosetting materials that were used for fabrication of glass fiber reinforced structural composites (42). The laminates contained 23-36% resin, were cured at 200-300°C and 1000-4000 psi, and gave flexural strengths of about 300,000 psi and moduli of 3 million psi.

Polymer-coated Electrodes

The use of modified vinyl polymers, with covalently or ionically bound organomettalic functions, has recently become a focal point of interest in the area of polymer-coated electrodes (43). Thus polyvinylferrocene, protonated poly(4-vinylpyridine)Fe(CN)$_6^{-3}$, the sulfonated fluoropolymer, Nafion, (modified with cobalt(II) tetraphenylporphyrin and more recently with both cobalt(II)tetraphenylporphyrin and Ru(NH$_3$)$^{3+}$ counterions and polymers of tris (4-vinyl-4'-methyl-2,2'-bipyridine)ruthenium(II), **17**, have been used. Such polymers act to protect the electrodes from oxygen or from a variety of corrosive reactions which occur when carrying out electrocatalysis or electrochemistry in surrounding solutions or when promoting hole/electron separation at semiconductor surfaces (43). They also act to catalyze oxidation or reduction processes in solution which do not occur at the semiconductor surface alone (10-13).

Metal Disposition

One class of vinyl polymers contains metals which are directly bonded via metal-carbon sigma bonds. These types are quite rare. Usually, transition metal-carbon bonds undergo rapid β-metal hydride elimination. However, this is not possible when β-hydrogens are absent. Pittman and Felix (44, 45) showed that the treatment of both linear and crosslinked chloromethylated polystyrenes with Mn(CO)$_5^-$, (η^5-C$_3$H$_5$)Mo(CO)$_3^-$ or (η^5-C$_5$H$_5$)W(CO)$_3^-$ resulted in displacement of chloride and formation of a stable metal-carbon bond within the polymers (i.e. **18** and **19**).

119

Thermal disposition of crosslinked resins containing these structures was carried out to see if the metal decomposition products could be dispersed within the polymer matrix. The decomposition of 18 released $Mn_2(CO)_{10}$ within the resin while 19 decomposed to produce the metal-metal dimer structure 20. The decomposition of metal carbonyls dissolved or dispersed in polymers is another aproach which has been reviewed by Goldberg (46).

Catalysis

Many orgaometallic monomers exhibit catalytic properties. Polymers containing such monomers may offer not only catalytic activity, but also certain stereoregulating properties as a consequence of the inherent steric requirements present about the metal-containing moiety within the polymer.

One of the most vigorously studied areas of application is the use of polymer supports for homogeneous catalysts (47-50). These polymer-anchored homogeneous catalysts can be used as fixed-catalyst beds in flow processes. The chemical selectivity inherent in the homogeneous organometallic catalyst is often maintained or in some cases enhanced (47-53). Other advantages of such bound catalysts include reduced reactor corrosion (associated with some homogeneous catalysts) (48), ease of product-catalyst separation, the ability to immobilize high concentrations of metal complex within the resin matrix thereby retarding bimolecular reactions which deactivate the catalyst (54-56), the ability to keep two mutually incompatible catalysts in a single reactor vessel for multistep synthesis (57-60), the ability to tailor the polymer-catalyst complexes to increase catalyst selectivity (61-64), and the ability to increase the reaction rate (62, 65-66). Some proprietary commercial processes are operating now using polymer-supported organometallic catalysts. Mobil Corporation took a polymer-supported rhodium hydroformylation process up to pilot plant scale in the mid-1970's before it was decided the projected markets would not justify new plant construction (67). Many patents have been issued covering the use of polymer-bound organometallic catalysts in hydrogenation (68, 69) hydroformylation (49, 69-74), hydroesterification of olefins (75), diene linear oligomerization (76), and other reactions (49).

The very interesting poly(silanes) have all silicon atoms in their backbones. The synthesis of soluble high molecular weight poly(silanes) was reported by West and coworkers (77-79) and by Wesson and Williams (80). Synthesis involves treating various dichlorosilanes with sodium at temperatures above 100°C. Homo- and copolymers are readily made. All of these polymers are strongly absorbing in the UV from 300 to 400 nm. UV irradiation of the poly(silanes) leads to molecular weight degradation. Silicon-silicon bonds are cleaved giving radicals. These radicals are useful in initiating polymerization reactions of monomers such as acrylates. Thus, small amounts of poly(silanes) act as photoinitiators for vinyl polymerizations where the wavelength of maximum initiation activity can be tuned by varying the substituents on silicon.

Ceramic Precursors

Organometallic polymers are being examined as precursors for ceramics. This area of research is very active. The expense in preparing nonoxide ceramics derives from the need for high purity precursors and the high cost in fabricating three-dimensional shapes at high temperature (> 1500°) often under high pressures (for example Si_3N_4, SiC and BN). At Nippon Carbon Co. polydimethylsilane is now a commercial precursor to silicon carbide fibers. Upon heating, polydimethylsilane is converted to a polycarbosilane with alternating Si and C atoms. This is fractionated and spun into fibers which are oxidized to the surface to make them rigid. Then pyrolysis at 800° to 1300° C gives silicon carbide fibers (81). Formable poly(silanes) can be converted to silicon carbide as fibers or coated films (82).

$$\left(\underset{\underset{CH_3}{|}}{\overset{\overset{CH_3}{|}}{Si}}\right)_n \xrightarrow{400°} \left(\underset{\underset{CH_3}{|}}{\overset{\overset{H}{|}}{Si}} - CH_2\right)_n \xrightarrow[\text{2. } 800°-1300°]{\text{1. oxidize surface}} SiC + CH_4 + H_2$$

Polymeric precursors to silicon nitride, Si_3N_4, are being actively sought. Seyferth treated CH_3SiHCl_2 with NH_3 to give complex oligomeric $[CH_3SiHNH]_n$ compositions which were dehydrocyclodimerized to sheet-like networks with functional composition averaging $[(CH_3SiHNH)_a(CH_3SiN)_b(CH_3SiHNK)_c]_n$. Eventually, these were pyrrolyzed under various conditions (including under NH_3) to give silicon nitride which was always mixed with some silicon carbide (83). Novel poly(silanes) have been prepared at Stanford Research Institute via dehydrocoupling of Si-H and N-H bonds. They were converted to Si_3N_4 in ceramic yields > 80% with purities of over 97% (84). Alternatively (86), ammonolysis of the adduct of SiH_2Cl_2 and pyridine in inert atmospheres gave poly(silanes) of the average structure $[(SiH_2NH)_{0.46}(SiH_2N)_{0.36}(SiH_3)_{0.18}]$.

$$Et_2SiH_2 + NH_3 \xrightarrow[60°]{Ru_3(CO)_{12}} \left(Et_2SiNH\right)_x + H_2$$
$$\underset{Si_3N_4 + SiC}{\xrightarrow{\Delta}}$$

Silicon carbide-titanium carbide composites have now been prepared by the pyrolysis of silicon-based polymers (87). Polymeric precursors to hexagonal boron nitride are also now beginning to attract attention although the availability of suitable boron-nitrogen polymers in the past has been lacking. The reaction of difunctionalized borazines with bistrimethylsilylamines results in the formation of soluble oils which readily form thin films (88). These films can be pyrolyzed to boron carbide/boron nitride ceramic coatings at 1200 where it was not yet possible to remove all of the carbon content (88).

REFERENCES

1. C. U. Pittman, Jr. and B. Suryanarayanan, <u>J. Amer. Chem. Soc.</u>, 96:7916 (1974).
2. C. Carraher, T. Manek, R. Linville, J. R. Taylor, L. Torre, and W. Venable, <u>Organic Coatings and Plastics Chemistry</u>, 44:753 (1981).
3. T. J. Marks, et al., <u>Abstracts 193rd ACS National Meeting</u>, Denver, CO, April 5-10, 1987, INOR 382; J. L. Petersen, C. S. Scharamm, D. R. Stojakovic, B. M. Hoffman, and T. J. Marks, <u>J. Amer. Chem. Soc.</u>, 99:286 (1977).
4. B. N. Diel, T. Inabe, H. K. Jaggi, J. W. Lyding, O. Schneider, M. Hanack, C. R. Kannewurf, T. J. Marks, and L. H. Schwartz, <u>J. Amer. Chem. Soc.</u>, 106:3207 (1984).
5. J. T. Wrobleski and D. B. Brown, <u>Inorg. Chem.</u>, 18(498):2738 (1979).
6. J. R. Reynolds, F. E. Karasz, C. P. Lillya, and J. C. W. Chien, <u>J. Chem. Soc.</u>, <u>Chem. Commun.</u>, 268 (1985).
7. H. Matsuda, H. Nakanishi, and M. Kato, <u>J. Polym. Sci. Polym. Lett. Ed.</u>, 22:107 (1984).
8. J. E. Sheats, C. Carraher, and C. U. Pittman, Jr., (eds.), "Metal-Containing Polymeric Systems", Plenum Press, New York, 1985.
9. M. Kaneko and A. Yamada, <u>Adv. Polym. Sci.</u>, 55 (1983) and ref. 5, pp. 249-274.
10. M. S. Wrighton (ed.), "Interfacial Photoprocesses: Energy Conversion and Synthesis", American Chemical Society, Washington (1980).
11. U. T. Muller-Westerhoff and A. I. Nazzal, <u>US Patent</u>, 4,379,740 (1983).
12. F. C. Anson et al., <u>J. Am. Chem. Soc.</u>, 106:59 (1984).
13. E. R. Savinova, A. I. Kokorin, A. P. Shepelin, A. V. Pashis, P. A. Zhdan, and V. N. Parmon, <u>J. Mol. Catal</u>, 32:149-159 (1985).
14. T. E. Bitterwolf in ref. 5, pp. 137-147.
15. B. J. Spalding, <u>Chemical Week</u>, Oct. 8, 29 (1986). C. M. Elliott et al., Abstracts, 193 ACS National Meeting, Denver, CO, April 5-10, 1987. INOR 232.
16. C. Carraher, S. Tsuji, J. DiNunzio and W. Feld., <u>Polymeric Materials</u>, 55:875 (1986).
17. C. Carraher and J. Schroeder, <u>Polymer P.</u>, 16:659 (1975) and <u>Polymer Letters</u>, 13:215 (1975).
18. L. Donaruma, <u>Polymer Preprints</u>, 22:1 (1981).
19. C. Carraher, R. Schwarz, J. Schroeder, and M. Schwarz, <u>J. Macromol. Sci. Chem.</u>, A15(5):773 (1981).
20. C. Carraher, V. Foster, R. Linville, and D. Stevenson, Abstracts 193rd National Meeting, Denver, CO, April 5-10, 1987, PMSE 95.
21. C. Pittman, <u>J. Paint Technol.</u>, 39:585 (1967).
22. C. Carraher and R. Dammeier, <u>Makromolekulare Chemie</u>, 135:107 (1970) and 141:251 (1971).
23. C. Carraher and R. Danmeier, <u>J. Polymer Sci.</u>, 8:3367 (1970) and 10:413 (1972).
24. C. Carraher and D. Winter, <u>Makromolekulare Chemie</u>, 141:237 (1971); 141:259 (1971); and 152:44 (1972).
25. C. Carraher and G. Scherubel, <u>J. Polymer Sci.</u>, 9:983 (1971).
26. C. Carraher and L. Wang, <u>Makromolekulare Chemie</u>, 152:43 (1972).
27. C. Carraher and J. Piersma, <u>J. Apl. Polymer Sci.</u>, 16:1851 (1972).
28. C. Carraher and J. Schroeder, <u>Polymer Preprints</u>, 16:659 (1971) and <u>Polymer Letters</u> 13:215 (1975).
29. E. J. Dyckman and J. A. Montemarano, <u>Am. Paint. J.</u>, 58(5):66 (1973).
30. A. N. Ghanem et al., <u>Eur. Polym. J.</u>, 15:823 (1979).

31. C. Carraher, D. Giron, D. Cerutis, T. Gehrke, S. Tsuji, R. Venkatachalam, and H. Blaxall, Org. Coat. Plast. Chem., 44:1 (1981).

32. H. Allcock, W. Cook, and D. Mack, Inorg. Chem., 4:2584 (1972).

33. H. Allcock, R. Allen, and J. O'Brien, J. Amer. Chem. Soc., 97:3914 (1977).

34. H. Allcock, in "Organometallic Polymers", C. Carraher, J. Sheats, and C. U. Pittman, Jr., (eds.), Chpt. 28, Academic Press, N.Y., 1978.

35. C. Carraher, D. Giron, W. Scott, and J. Schroeder, J. Macromol. Sci. Chem., A15(4):625 (1981).

36. C. Carraher, D. Giron, I. Lopez, D. R. Cerutis, and W. Scott, Organic Coatings and Plastics Chemistry, 44 (1981).

37. B. Block, Inorg. Macromol. Chem. 12:115 (1970).

38. N. E. Wolff and R. J. Pressley, Apply. Phys. Lett., 2:152 (1963).

39. Y. Okamoto, S. S. Wang, K. H. Zhu, E. Banks, B. Garetz, and E. K. Murphy, in: ref. 5, pp. 425.

40. J. E. Sheats, F. Hessel, L. Tsarouhas, K. G. Podejko, T. Porter, L. B. Kool, and R. L. Nolan, in: "New and Unusual Monomers and Polymers", B. M. Culbertson and C. U. Pittman, Jr. (eds.), Plenum Press, NY, pp. 83-98 (1983); Polymeric Mat. Scil and Eng., 49(2):363 (1983); also in ref. 15, pp. 83-98.

41. C. Carraher, Chemtech., 744 (1972).

42. N. Bilow and H. Rosenberg, J. Polymer Sci., A-1(7):2689 (1969); N. Bilow and H. Rosenberg, U.S. Pat. 3, 640, 961 (1972).

43. L. R. Faulkner, Chem. Eng. News, 28, Feb. 17, 1984.

44. C. U. Pittman, Jr. and R. F. Felix, J. Organometal. Chem., 72:389 (1974).

45. C. U. Pittman, Jr. and R. F. Felix, J. Organometal. Chem., 72:399 (1974).

46. R. Tannenbaum, E. P. Goldberg, and C. L. Flenniken, "Metal-Containing Polymer Systems", J. Sheats, C. Carraher, and C. U. Pittman, Jr., (eds.), Plenum Press, New York, p. 303-340 (1985).

47. C. U. Pittman, Jr., in: "Comprehensive Organometallic Chemistry", G. Wilkinson, F. G. A. Stone, and E. W. Abel (eds.), Permagon Press, Oxford, Vol. 8, Chapter 55, 553-608 (1982).

48. F. R. Hartley, "Supported Metal Complexes", D. Reidel Publishers, pp. 1-318 (1985).

49. Y. Chauvin, D. Commereuc, and F. Dawans, Prog. Polymer Sci., 5:95 (1977).

50. C. U. Pittman, Jr. and G. O. Evans, Chemtech 3:560 (1973).

51. C. U. Pittman, Jr., in: "Polymer-Supported Reactions in Organic Synthesis", P. Hodge and D. C. Sherrington (eds.), Wiley, Chichester, pp. 249-291 (1980).

52. N. Toshima, Yuki Gosei Kagaku Kyokaishi, 36:909 (1978), Chem. Abstr., 90:104033 (1979).

53. V. M. Akhmedov, A. A. Medzhidob, and A. G. Azizov, Aserb. Khim. Zh., 122 (1979), Chem. Abstr., 92:11692 (1980).

54. C. U. Pittman, Jr. and Q. Ng, J. Organometal. Chem., 153:85 (1978).

55. W. O. Haag and D. D. Whitehurst, in: "Catalysis", J. W. Hightower, ed., Vol. 1, 465 (1973).

56. R. H. Grubbs, C. Gibbons, L. C. Kroll, W. D. Bonds, Jr., and C. H. Brubaker, J. Amer. Chem. Soc., 95:2374 (1973) and 97:2128 (1975).

57. C. U. Pittman, Jr. and L. R. Smith, J. Amer. Chem. Soc., 97:1749 (1975).

58. C. U. Pittman, Jr., L. R. Smith, and R. M. Hanes, J. Amer. Chem. Soc., 97:1742 (1975).

59. R. F. Batchelder, B. C. Gates, and F. P. J. Kuijpers, Preprint A 40, Sixth International Congress on Catalysis, London (1976).

60. D. E. Bergbreiter and R. Chandran, J. Am. Chem. Soc., 107:4792 (1985).

61. R. H. Grubbs, L. C. Kroll, and E. M. Sweet, J. Macromol. Sci., A7:1047 (1973).
62. C. U. Pittman, Jr. and R. M. Hanes, J. Amer. Chem. Soc., 98:5402 (1976).
63. C. U. Pittman, Jr. and A. Hirao, J. Org. Chem., 43:640 (1978).
64. V. A. Kabanov, V. G. Popov, V. I. Smetanyuk, and L. P. Kalinina, Vysokomol. Soedin, B23:368 (1981).
65. S. Jacobson, W. Clements, H. Hiramoto and C. U. Pittman, Jr., J. Mol. Catal., 1:73 (1975).
66. R. H. Grubbs, C. P. Lau, R. Cukier, and C. H. Brubaker, Jr., J. Am. Chem. Soc., 99:4517 (1977).
67. W. H. Lang, A. T. Jurewicz, W. O. Hagg, D. D. Whitehurst, and L. D. Rollman, J. Organometal. Chem., 134:85 (1977).
68. N. L. Holy, W. A. Logan, and K. D. Stein, US Patent, 4,313,018 (1982).
69. Mobil Corp., Belgian Patent, 721,686 (1969).
70. W. O. Haag and D. D. Whitehurst, Mobil Corp. US Patent 4,098,727, (1978).
71. A. J. Moffat, J. Catal., 18:193 (1970) and 19:322 (1970).
72. K. G. Allum and R. D. Hancock, British Patents, 1,277,737 and 1,295,673 to British Petroleum (1972).
73. W. O. Haag and D. D. Whitehurst, US Patent 4,098,727 to Mobil Corp. (1978).
74. H. B. Tinker and D. E. Morris, US Patent 4,052,461 to Monsanto Inc. (1977).
75. C. U. Pittman, Jr. and Q. Y. Ng, US Patent 4,258,206 to University of Alabama (1981).
76. C. U. Pittman, Jr. and R. M. Hanes, US Patent 4,243,829 to the University of Alabama (1981).
77. K. S. Mazdyasni, R. West, and L. D. David, J. Am. Ceram. Soc., 61:504 (1978).
78. R. West et al., J. Am. Chem. Soc., 103:7352 (1981).
79. R. West, J. Organometal. Chem., 300:327 (1986).
80. J. D. Wesson and T. C. Williams, J. Polym. Sci. Polym. Chem. Ed., 18:959 (1980).
81. S. Yajima, H. Hayashi, and M. Omori, Chem. Lett. 931 (1975);
    S. Yajima, K. Okamura, and J. Hayashi, Chem. Lett., 1209 (1975).
82. R. West et al., Am. Ceram. Soc. Bull., 62:825 (1983).
83. D. Seyferth and G. H. Wiseman, J. Am. Ceram. Soc., 67:C-132 (1984); US Patent 4,482,669 (13 Nov. 1984).
84. Y. D. Blum, R. M. Laine, K. B. Schwartz, and D. J. Rowecliffe, Patent Pending.
85. R. M. Laine, Y. D. Blum, A. Chow, R. Hamlin, K. B. Schwartz, and D. J. Rowecliffe, Polymer Preprints 28(1):393 (1987).
86. M. Arai, S. Sakurada, T. Isoda, T. Tomizawa, Polymer Preprints, 28(1):407 (1987).
87. S. Yajima et al., J. Materials Sci., 16:1349 (1981).
88. C. K. Narula, R. T. Paine, and R. Schaeffer, Polymer Preprints, 28(1):454 28(1):454 (1987).

POLYMERIC FOAMS

Raymond B. Seymour

Department of Polymer Science
University of Southern Mississippi
Hattiesburg, Mississippi 39406-0076

Natural sponges (<u>Phylum</u> <u>Porfera</u>) are organless multicellular porous species which have been used for centuries as water absorbants, such as bath sponges. Wood and straw are also cellular materials and cork and balsa wood have also enjoyed a ready market because of their insulating properties and their high strength to weight ratio.

Hence, when polymeric synthetic foams became available, there was a ready market for these versatile products with low specific gravity. The first synthetic cellular plastic was produced accidentally by Leo Baekeland in 1909 but the first commercial foam was sponge rubber, which was produced by heat curing rubber in the presence of sodium bicarbonate.

The cells are interconnected in opened celled (multicellular) foams and discreet in closed (unicellular) foams. Foams may be elastic, like sponge rubber or rigid, like Baekeland's phenolic foam.

The properties of a foam are related to structural variables, such as composition, geometry and specific gravity. The compressive strength of a cellular plastic is dependent on composition and specific gravity. The latter is essentially independent of temperature but physical properties usually decrease as the temperature is increased.

Insulating ability of a foam is related inversely to the thermal conductivity through the solid phase. However, the thermal conductivity of the air-propellant mixture in the cells must also be considered. These values vary from 0.0085 to 0.00259 W/m.k for trichlorofluoromethane and air, respectively.

The propellant which causes expansion of the polymer may be the result of gas formation resulting from the thermal breakdown of a component of the system, the presence of an additive gas or a highly volatile liquid. Rubber latex foam has been produced by whipping air into the latex; polyurethane foam may be made by adding moisture or carboxylic acids to the reactants and some foams are produced by the addition of steam.

A different type of foam, called a syntactic foam, is produced by the incorporation of hollow beads in the resin. Cellular products have a relatively strong integral skin have improved strength and are called structural plastic foams.

Polystyrene foam (Styrofoam) is produced by injecting a volatile liquid, such as pentane, into the barrel during the extrusion of polystyrene. Injection molded polystyrene foam may be produced in a similar manner by adding the expanding agent to the barrel before the plunger forces the mixture into the mold cavity. In an alternate technique, polystyrene beads are produced by suspension polymerization, in the presence of a volatile liquid, such as pentane. The composite beads are molded after being thermally expanded by steam. It is customary to add a nucleating agent, such as talc, to the molten polymer to assure uniform foam production.

Polymeric foams may also be produced by casting or spraying the reactants in the presence of an expanding agent. When this process occurs in a split mold, it is called reaction injection molding (RIM).

The thermal conductivity (K_{foam}) of a foam may be calculated from the thermal conductivity of the gas (K_{gas}) using the following formula, in which the density is stated as lb/ft^3 and the conductivity of the foam is expressed as a K value in $BTU-in/FT^2-F^o-hr$).

$$K_{foam} = K_{gas} + 0.018 + 0.02D$$

Foams based on expanded urea-formaldehyde resin have been produced, in situ, in walls of residential dwellings. However, because of the possibility of the production of formaldehyde by thermal or hydrolytic degradation, the use of these foams has declined. Polystyrene continues to be the most widely used rigid polymeric foam.

Foams have also been produced from polyolefins, PVC, cellulose acetate, phenolic, epoxy and silicone resins and polyurethanes. The latter, which was discovered accidentally by Dr. O. Baeyer, are now widely used as both rigid and flexible foams. The annual consumption of polymeric foams, in thousands of tons, in the USA is as follows:

polystyrene	535
polyurethane flexible foam	650
polyurethane rigid foam	350
PVC foam	310

The principal uses for polymeric foams are as follows: insulation, cushioning, packaging, construction, transportation, consumer use, furniture, flooring, bedding and appliances.

USE OF AN EFFECTIVE QUALITY MANAGEMENT SYSTEM

FOR BUSINESS IMPROVEMENT IN THE PLASTICS INDUSTRY

Ralph A. Hovermale

High Performance Films
E.I. du Pont de Nemours & Co., Inc.
Circleville, Ohio

INTRODUCTION

From the discovery of cellophane by the French and the fundamental polymer work with Nylon by Carothers in the early 1930's, a tremendous market has developed for plastics in all areas of consumer's lives. Plastic pellets and powders are molded into all manner of shapes such as pipes, toys, household goods, automotive parts, apparel items and the like. Plastic films are found in such items as video tapes, audio tapes, capacitors, motors, packaging applications, and weather balloons. Plastic fibers are predominant in apparel, carpets, linens, draperies and similar industries. From a simple beginning in the first quarter of the twentieth century, the plastics industry has grown to be an indispensable giant on a worldwide basis. As markets have increased, the competition to supply these markets has also increased. The customer has become more knowledgeable and, with a wealth of offerings to choose from, has become very selective. He demands from his supplier a sophisticated total offering which must include the best combination of price, quality, and service. The suppliers that meet this requirement are successful while those that do not find their products and services in little demand. One of the most successful tools for use by suppliers in meeting these demands has been a strong, broadly based, integrated Quality Management System with an emphasis on statistical process control and strong supplier-customer quality partnerships.

THE QUALITY SYSTEMS MODEL

A three-fold model of a good Quality Management System is composed of the Quality Cycle, Generic Concepts for Suppliers and Customers, and Continuous Improvement. These items form a base for understanding of basic principles and the movement of information and action within the system.

The Quality Cycle

The cycle for the development of quality of a product or service between suppliers and customers is shown schematically in Figure 1.

<u>Fitness for Use</u>. The customer has a definite need from the supplier described as "Fitness for Use". A material, "fit for use", will, when used by the customer, deliver a high level of satisfaction to the customer's customer. It will provide a good yield and profit margin to the customer and will be converted by his organization with a minimum of problems at an acceptable high throughput.

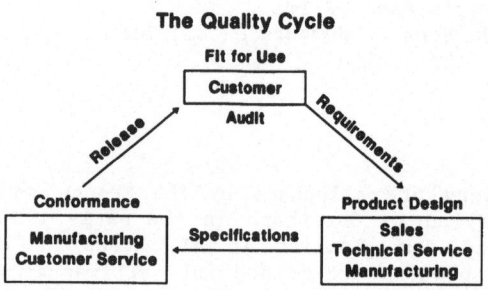

Figure 1. The Quality Cycle

<u>Requirements</u>. The term "Fitness for Use" is described by the customer in terms of his final product and his business. This concept must be reduced to "Requirements", which describes the product provided by the supplier and, when stated in terms of the product produced by the supplier, becomes the "Specification". It will probably contain an average and acceptable range for each specified property, and may possibly require use of a standard process and standard procedures, along with standard procedures and quality assurance for the supplier's suppliers. The transformation of "Fitness for Use" from the customer's language to "Specifications" in the supplier's language ("Product Design") is a key step which will influence, to a large degree, the satisfaction of the customer with the supplier's product and may also impact greatly on the supplier's earnings as associated with production cost of this material. Normally, the product specification step is coordinated by the supplier's Technical Marketing representatives with prime participation from his Sales and Manufacturing groups.

Conformance to Requirements. Once the "Requirements" are established, it becomes the responsibility of the Manufacturing organization to conform to these requirements with the product it ships to the customer. This is approximated by use of a "Product Release System" which guarantees conformance to the "Specifications" for all products shipped to the customer.

Audit. The final audit is conducted by the customer actually using the material or service in his plant. If it proves satisfactory, specification and release lock its quality into place and "Control" is established. If the material or service is not satisfactory, the Quality Cycle is continued through other cycles, as necessary, until the material or service is acceptable, then "Control" is established.

Generic Concepts for Supplier - Customer Relations

All businesses and individuals have both suppliers and customers. Each is also both a supplier and a customer, depending on the particular material or service in question. Figure 1 is a model for either situation. As a customer, the business or individual places himself in the customer block in the figure and views the cycle from that point. As a supplier, he places himself in the other two blocks and views the cycle from these points. It is, however, always the same cycle.

Since each business or individual is both a supplier and a customer, certain generic concepts must apply to Supplier-Customer Relations. Such a set of concepts is shown in Table I.

The underlying themes of these concepts are the idea of a quality partnership involving mutual risk and mutual gain, making a product right the first time, and use of statistical control and release tools. These concepts provide a large part of the base of any valid Quality Management System.

Table I. Generic Concepts for Supplier-Customer Relations

Generic Concepts for Supplier-Customer Relations

- A good supplier-customer relationship is one in which both parties are aware of, and in reasonable compliance with, the other's philosophies, requirements and expectations. They form a Quality Partnership.

- These partnerships work toward the mutual benefits of both parties. They are responsive and they share and communicate.

- A few top quality suppliers with competitive total offerings will become the long-term partners.

- Customers and suppliers form two links in a continuing chain.

- The best quality product and most cost-effective process are obtained when the product is made right the first time.

- Proper use of statistical techniques and concepts offers the most effective means of accomplishing process control, measurement control, and product release.

- Incoming material testing by the customer is not cost-effective. Testing by the supplier partner is the most desirable procedure.

Continuous Improvement

Once the Quality Cycle is established and a satisfactory product is being routinely produced, the Improvement Process must be instituted. Usually, the customer-partner and his products are under attack from competition, both from the quality/service and from the cost standpoints. The supplier-partner is under a similar attack from his competition and the ultimate consumer is becoming more sophisticated and demanding. The need for continuous improvement is a fact of business survival. Figure 2 shows a diagram of an unending cycle for continuous improvement.

Control. In the "Control" phase, a "Specification" which correctly matches today's "Requirements" is in place. A control element, preferably statistically based, which allows for the "building in" of quality is operating. By "building in", one means that deviations of the product or process from goal values are detected immediately when statistically significant, and instant adjustment is made to return the product or process to goal values. Finally, a "Release" element, preferably statistically based, is in place to assure Conformance to the "Requirements".

Improvement. After "Control" is established, the Quality Cycle will continually produce a product that is consistent, meets conformance, and is fit for use until the customer's need changes. A good supplier-partner will work with the customer-partners to define those properties, services, new products, prices, and general offerings that will be needed in the future. In addition, he will investigate internally to define those areas for appreciable improvement in costs, consistency, and productivity which would result in higher profitability and better leverage for his business. These investigations, both externally and internally, lead to the establishment of projects. According to Juran[1], a project is a problem scheduled for solution. Projects come from many sources such as Cost of Quality Analysis, customer surveys, feedback from Sales, field surveys, surveys of competition, technical and research studies, business analysis, long range corporate planning, and chronic problem areas. Projects are analyzed for worth and timing and the highest priority items are taken as projects for immediate attention. Assignment of resources is made and work begins. The chronology of events normally taking place is best described by Juran in his book, "Managerial Breakthrough"[2]. This concept is shown graphically in Figure 3.

Continual Improvement

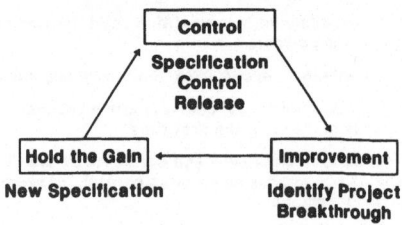

Figure 2. Continual Improvement

In this example, yield loss is shown on the ordinate and elapsed time on the abscissa. Examination of the left hand portion of the curve reveals two primary features. The curve is relatively flat, with short term variations, except for one large digression. The flat portion represents a relatively constant level of chronic yield loss that the manufacturer has become accustomed to enduring. The short term variations represent the small daily, weekly, or monthly perturbations that are controlled by the current state of knowledge of operating personnel and maintenance of the process and equipment. The large digression represents a sporadic problem or "rare event" which requires the generation of new knowledge or approaches and is quite frequently handled by a special task team. These "rare events" will occasionally occur and will be dealt with accordingly.

"Breakthrough" Concept

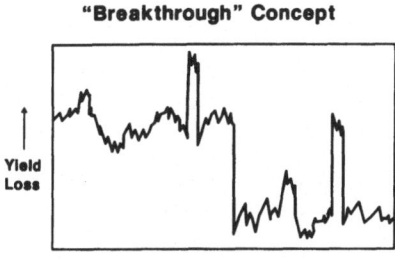

Figure 3. "Breakthrough"

It is the chronic problem area where great improvements can be made. With assignment of resources to the highest priority projects, solutions can be accomplished with the result a major improvement in yield loss as shown in the right hand portion of the curve in Figure 3. Note that the sporadic digression still occurs. The large step change in going from the higher level of chronic yield loss in the left hand portion to the lower level in the right hand portion is called "Breakthrough". A similar diagram may be drawn for each project with the time scale changing and the ordinate relabeled, as appropriate, to show the problem and the improvement. Labels can be such typical items as share loss/gain, customer complaints/returns, or operating downtime.

Hold the Gain. All too frequently, the improvement made in the "Breakthrough" step is not institutionalized and the gain is lost as the total process slides back to the old level of performance. In order to "Hold the Gain", it is necessary to draft the gain as a "New Specification" which then takes the place of "Specification" in the "Control" step. The process is now required to be controlled to the new level of performance. The "Release" step is also adjusted to the new level to insure all goods and services conform to the new specification.

These models have been evaluated in actual field trials. The following examples are taken primarily from work performed with High Performance Films in the Du Pont Company between 1982 and 1986.

Elements of a Quality Management System

A good Quality Management System contains the elements given in Table II as a minimum. These elements may be considered individually with specific examples where appropriate.

Statistical Data Base - Design and Maintenance. Establishment of a means of collecting, analyzing, and reporting data is an initial step. For extremely small installations, hand documentation with assistance from a pocket or desk calculator may suffice. For small to intermediate facilities, a system based on personal computers currently available may be appropriate. However, for large plants or for large plants networked onto a single computer system, a large mainframe computer system will be necessary.

Once the computer system is established, a next step is to classify properties to be controlled and released. The most important to the customer and hardest to control by the supplier will be given the top priority and the highest level treatment. Those of least importance to the customer or those which are constant in the suppliers process are given less priority and lower level treatment. It should be noted that, if 2 or more properties exist that move in unison to the same process knobs, then the more important may receive the higher priority approach and the less important may receive the lower priority approach.

The sampling plan should be reviewed to determine whether it is in line with the property classification. The higher priority items should be sampled frequently while the lower priority items probably need to be sampled less often or merely monitored on a daily or weekly basis. For the higher priority items, the sampling program should be expanded to allow separation of laboratory variability from product variability.

Table II. Elements of the Quality Management System

Elements of the Quality Management System

- Design and maintenance of statistical data base
- Method control
- Standard operating system
- Process control
- Statistical consideration - ANOVA, release, problem analysis
- Specifications
- Release
- Supplier relations
- Cost of quality
- Customer relations
- Continual improvement
- Audit

<u>Method Control</u>. After establishing a sampling plan and data base, attention is turned to the laboratory. If a Quality Management System is to be established, laboratory results must be completely reliable. It is not sufficient to possess a method and to calibrate periodically against a standard. Knowledge of the variability of the test (including all operators, all shifts, all equipment, and all seasons) is necessary in order to separate test variability from product variability. An on-going control charting system is needed to detect statistically significant shifts from goal or "Aim" value for the control sample. Resampling must be strictly forbidden unless there is a definite known assignable cause for an "Off-Aim" reading (incorrect sample taken, sample accidentally contaminated, etc.). If more than one facility is involved, frequent Interlaboratory Checks (ILC) must be carried out for important properties over a range of values. This involves Plant Laboratories, Sales Laboratories, and Research Laboratories. The best system is probably one in which all laboratories are periodically furnished with sets of control samples from the same control material so that their daily method control systems are operated using the same material at each laboratory. An on-going IMC (Interlaboratory Method Check) is the result. A regular ILC run semi-annually is sufficient to audit the on-going IMC.

Control charting falls into two general categories. The first, based on techniques developed by Shewhardt[3], involves detection of product drift outside a preset limit (usually ±3 or ±4 standard deviations). After detection, the process is adjusted to bring the drift back within limits. This is a good technique for large deviations from Aim. The second category is the CUSUM[4] technique which detects statistically significant product drift Off-Aim. After detection, the process is adjusted to bring the drift back to Aim. This is a good technique for smaller deviations from Aim and will be discussed with examples in this chapter.

It is essential that all participating laboratories use the same documented methods and, as far as possible, the same equipment. Replacement or rebuilding of equipment requires requalification with the standard sample. (Usually 10 or more data points are sufficient for a good approximation of the mean and 35 for the standard deviation. An update every 60 to 90 data points is highly recommended.)

The elements of Method Control can be summarized as shown in Table III.

Table III. Elements of Method Control

Method Control

- Standard equipment
- Standard method
- Cusum control
- Interlaboratory method checks
- Interlaboratory checks

Table IV. Improvement in Test Variability by Use of Method Control

Examples of Improvement in Variability (%)

Property	Plant 1	Plant 2
Tensile	36	44
Modulus	46	66
Clarity	12	70
Haze	37	86
Shrinkage	80	67
Color	57	82

The effect of Method Control in improving test variability in the first 6 months of use is shown in Table IV. These data show a remarkable improvement for two plant laboratories at plants some 500 miles apart using the same control sample.

An unexpected result can be that several laboratories, using the same method, same equipment, and measuring the same standard samples can have different results. In Table V, several sets of average (Aim) values for various properties are shown which were measured at 3 separate plants using identical control samples. Several hundred separate measurements are involved. Information such as this can lead to productive method trouble shooting and demonstrates vividly the absolute need for ILC's between customers and suppliers and a good reason for only the supplier partner measuring the product with a method under good statistical control.

Another advantage of Method Control is the detection of tests with little validity. In Figure 4, three time series plots are shown for a common tensile property control sample measured at three different plants. It can be noted that, once again, the plants obtain different values for the property. However, each notes a distinct downward curvature of the value in the summer months. This chart was repeated

Table V. Bias in Test Methods

Test Method Control

Test	Aim Value - Standard Sample		
	Plant 1	Plant 2	Plant 3
Tensile	26.8	26.7	29.0
Modulus	582	595	645
Elongation	119	115	128
Shrinkage	1.62	1.67	1.86
Clarity	52.9	54.1	53.5
Haze	20.3	22.4	24.3

Interlaboratory Check
Tensile Property

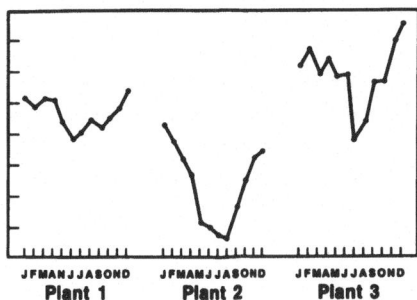

Figure 4. Unstable Test Method

over an additional 18 month time period with the same results. This was a prime property for process control and release. With the seasonal variability, there was a great deal of doubt that the product produced during summer months was the same as that produced during winter months. The process was actually adjusted to a different level causing problems with other properties. As a result of this work, this property was abandoned for control and release and was replaced by another, more stable property which accurately reflected the polymer structure desired by the customer.

Process Control. After Method Control is established, Process Control follows the same general steps. Standard equipment is designated and is not changed or replaced without requalification. Standard Operating Conditions are set with allowable adjustment or control limits. An adjustment outside of the standard limits is reason for stock restriction because of a non-standard process. It is also very important that all personnel follow the same procedures in operation and that these procedures are well thought out and documented. This set of procedures is called the Standard Operating Procedures. Finally, the product is maintained On-Aim by Process CUSUM Control. A summary of these elements is given in Table VI. An example of the improvement in product variability that can be gained from Process Control is shown in Figure 5 for shrinkage of a polymer sheet at elevated temperatures. The first 34 data points represent monthly averages for 34 months of production before Process Control was invoked. The next 23 points represent the monthly averages for the first 23 months after Process Control was established.

A further sophistication in CUSUM design may be undertaken after the Method and Process CUSUMS are well defined. This involves "CUSUM Matching" which demands designs such that an Off-Aim Method will trigger a Method alarm long before it will trigger a Process alarm and cause unnecessary process adjustments.

Table VI. Elements of Process Control

Process Control

- Standard equipment
- Standard operating conditions
- Standard operating procedures
- Cusum control

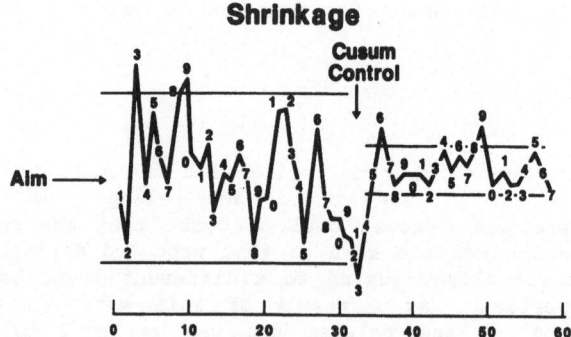

Figure 5. Process Improvement by "On-Aim" Control

<u>Statistical Analysis, Specifications, and Release</u>. The relationship
between "Specifications" and "Release Limits" are shown schematically in
Figure 6.   "Specifications" may be seen as based on true product
variability without lab error, while "Release Limits" include both
product and laboratory measurement variability which is usually referred
to as an Observed Value. Laboratories measure and report Observed Values.
By proper statistical treatment of the information gathered from the
sampling plans and placed in the data files mentioned earlier, the True
Values may be separated from the Observed Values.   This step allows one
to determine the large source of variability and, if appropriate, set
about an improvement project to reduce it.   By proper programming, the
data may be analyzed to provide the best (lowest cost) combination of
sampling plan and release limits that continue to provide conformance to
specifications.

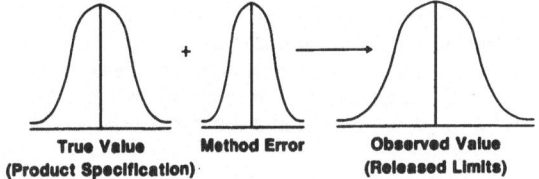

Observed vs. True Values

True Value	Method Error	Observed Value
(Product Specification)		(Released Limits)

Figure 6.   Specifications vs. Release Limits

Once "Specifications" are stated in terms of "True Values", the product may be transferred to other manufacturing facilities (with different laboratory variability or Method Aim values) and still be matched to the original product. Also, if the laboratory variability changes in a given plant (as measured on periodic updates), a simple change in Release Limits is all that is needed to continue to produce product to the same specifications.

<u>Segregated Product and Release</u>. There are times that a product may meet Release Limits but is not a representative product from the normal product distribution. This product is not released normally but is segregated or set aside for special disposition by the Marketing groups. Such situations are summarized in Table VII by cause.

Table VII.   Product Segregations

Non-Steady State Process
    1.   Start Ups
    2.   Transitions
    3.   Short Runs

Non-Standard Process
    1.   Outside Standard Conditions
    2.   Improper or Unqualified Procedures Used

Lack of Conformance to Specification
    1.   Off Aim - too far/too long

The non-steady state process situations are obvious. Product made under these conditions is not guaranteed to fall within the normal product distribution or to represent normal product and as a consequence represents a risk that must be removed before passing on to the customer. When a material is made by a non-standard process or procedure, there is no guarantee that some unmeasured property, which helps determine "Fit for Use", may vary and cause a customer problem. The customer must not be exposed to this risk. Finally, a substantial amount of product produced Off-Aim for long lengths of time offers a risk that, although it may be within specifications, its distributor is centered at a different point than normal and may cause problems to the user. Product falling in any of these categories is to be segregated for further testing, sale in non-critical areas, more restrictive release limits, etc. Normally, less than 5% of the product made falls within the segregated category. A value of less than 2% is met by the better producers.

In order to maintain a good cost situation for the supplier-partner, the segregated product must be sold at its maximum value. However, minimum risk must be placed on the customer-partner. He must continually receive a product which meets the written specifications and is "Fit for Use". A schematic design of a Release System which meets these criteria is shown in Figure 7. The procedure shown in this schematic representation allows first for segregation of non-steady state material. Steady state material is processed through normal lot release and, if it passes, results in one product good for all customers. Segregated material must be analyzed as to its projected "Conformance to Requirements". This may involve such items as 100% restriction of the product, additional testing, special consultation with Marketing, or discussions with selected customer-partners. The sample system shown in Figure 7 requires the application of tighter release limits, which, if met, allow shipment to all accounts. The material that passes neither the segregated release nor the lot release steps forms a miscellaneous grouping of materials that must be released only to customers who place low or no value on the property restricted, customers who prefer material distributions with some skew favoring the property value under

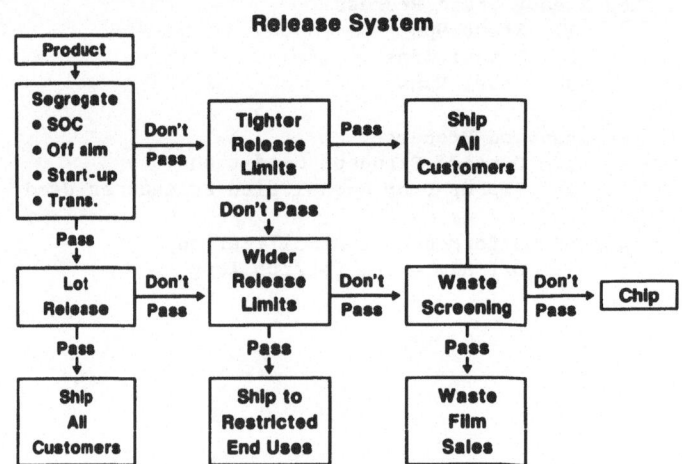

Figure 7. A Segregated Release System

consideration or customers paying a lower price for this material. This grouping of miscellaneous materials is referred to as "Wider Release Limits" in Figure 7. Material passing the requirements is shipped to "Restricted End Uses". Material not passing falls into a waste category out of which material is either recycled (or destroyed), reworked, or sold into a market with little or no quality demand on the properties involved for the price paid. Although highly dependent on the industry, many commodity products with a reasonable Quality Management System in place might be expected to produce a distribution breakdown as follows:

Product	100%
Pass Segregation	>95%
Pass Lot Release	>90%
Pass Tighter Release	> 3%
Pass Wider Release	> 4%
Waste	< 1%
Recycle	< 2%

For the speciality or highly technical industries, the amount of material passing the Segregation, Lot Release, Tighter and Wider Release steps is usually considerably lower and the waste is much higher. An effective Quality Management System in these areas usually offers great opportunities for cost and quality improvements to the supplier-customer partnership.

Supplier Relations. With the in-house "Control" phase implemented (Specification through Release), the supplier-partner must continue back on the Quality Cycle and assume the customer-partner role with his suppliers. A medium sized company may spend 1 to 2 billion dollars a year on essential materials and services. The same overall generic quality system is anticipated to be developed by the suppliers of goods and services. The suppliers are expected, at a minimum, to take the following specific actions.

1. Have in place an effective Quality Management System. A statistically based system is preferred.

2. Have well documented sampling plans and test methods in use and communicate them to the customer-partner.

3. Provide certification of analysis or on-going process data to allow the customer-partner to monitor performance and correlate with performance at his plant.

4. Provide prior notification to the customer-partner regarding any significant changes to the process equipment or conditions, raw materials, sampling plan or test methods in sufficient time for the customer-partner to analyze the effect on his process/product. A test run may be required.

The "Partnership" concept is stressed in the relationship resulting in meaningful, complete, and mutually agreeable specifications, data sharing, and a joint path forward to common goals and shared opportunities.

The customer-partner provides an effective audit system, agreeable to the supplier-partner, to insure that the Quality Cycle and Continuous Improvement are effective and on-going.

<u>Customer Relations</u>. The effective supplier gladly enters into the Quality Cycle. He is a willing partner and works to define realistic expectations by the customer-partner. He then meets those expectations. He strives to be proactive in information and data sharing. He agrees to a mutual path forward for the partnership and aggressively meets his commitments to it. He assists the customer-partner in a friendly, cooperative manner in auditing the effectiveness of this Quality Cycle.

GAINS

The gains from use of a good Quality Management System are numerous. Some of the most noteworthy may be discussed in more detail.

<u>Product Consistency</u>

As was seen in the discussion under the Control section, product consistency is improved several fold by application of the control techniques to method, process, and raw materials.

<u>Productivity</u>

Improved consistency and a product "made right the first time" leads to higher first pass yields and higher productivity. There have been several cases noted of new production records being set when a plant has commenced operating under a vigorous Quality Management System.

<u>Cost</u>

As productivity increases and first pass yields increase, the need of rework and recycling manpower and equipment is reduced. Also, because of improved product consistency and product design, customer complaints and returns are reduced. The result is a lower cost product. It is interesting to note that many of the production records mentioned in the productivity discussion were accomplished at or near record low costs! A good example of this process, which is also a "Continued Improvement" process, is shown in Figure 8. The first 24 points represent 24 months of data gathered for a critical tensile property for a product for the packaging industry. The product was poorly accepted in the field. In an

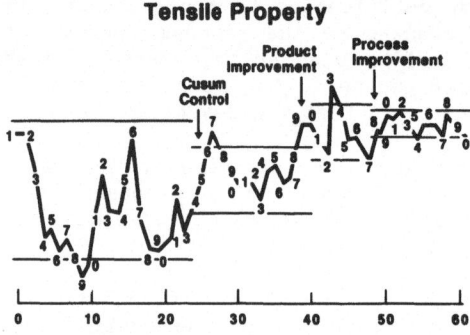

Figure 8. Continual Improvement

effort to avoid the high and low dispersions of the process, customers had demanded special specifications which resulted in some 15 separate products. These were obtained by the poor technique of sorting through the production and separating it into the 15 categories. There was no attempt to make one product good for everyone or to install feedback control loops to "build quality in". The result was a non-competitive product, a complicated, ineffective release system, numerous complaints and returns, and poor productivity and profitability. At the 25th data point, CUSUM control and a standard operating system were begun. Results were dramatic! The extreme high and low digressions were eliminated, a more disciplined release system was started, and a change in production philosophy was initiated. After a time (15 data points or months), yields began to increase, less recycled product was available for inclusion in the polymers, and improvements were made in the operation of the process. It became impossible to maintain the tensile property On-Aim with these improvements. A review was held between Manufacturing and Marketing. Two alternatives were available.

1. Produce a product at a higher Aim value (improved product) at the same cost.

2. Continue producing the same product but change the process to a lower cost.

Marketing's discussion with customers led to an immediate choice of the first path. The result is shown in the next 9 data points (months). At this level of operation, and at the same cost, remarkable changes occurred. The product was now competitive, the 15 separate customer specifications disappeared. One product, good for everyone, was produced, customer complaints and returns were practically non-existent, and the reputation of the supplier-partner was greatly enhanced. By the 49th data point from the beginning, another change was evident. Yields had again increased. The property could no longer be maintained at the lower level. A choice must be made again. This time, Marketing chose to go the route of lower cost while maintaining the Aim. The final 12 points show results after this decision. As an added benefit, the plant had lowered the variability even further. The overall result of following this continuing improvement cycle is that now a competitive product with excellent consistency and well accepted by customers is being produced at a lower cost! Two production records were also set during the course of the last 12 data points. An excellent discussion of concepts which agree well with this cycle is provided by Crobsy[5].

Customer Satisfaction

The consistency, product design, and cost accomplishments are merged into the total offering to the customer-partner of quality, price, and service to provide the base for improved share for the supplier-partner and improved earnings for both partners.

Profitability

The combination of higher production, lower cost, and increased market share yields greatly improved profits for the supplier-partner. The combination of product design, consistency, price, and service yields greatly improved profits for the customer-partner. Mutual goals, a common path forward, and a cooperative working relationship give both partners a competitive position for continued mutual benefits.

REFERENCES

1. J.M. Juran and Frank M. Gryna, Jr., "Quality Planning and Analysis",
        2nd Edition, McGraw-Hill Book Company, New York, New York (1980)
2. J.M. Juran, "Managerial Breakthrough", McGraw-Hill Book Company,
        New York, New York (1964)
3. W.A. Shewhardt, "Economic Control of Quality of Manufactured Product",
        Van Nostrand Publishing Company, Princeton, New Jersey (1931)
4. BS 5703, <u>Data Analysis and Quality Control Using CUSUM Techniques
        Part I</u>, British Standard Institute, London (1980)
5. P.B. Crobsy, "Quality is Free", The New American Library, New York,
        New York (1979)

APPLIED POLYMER SCIENCE - A FORECAST

Herman F. Mark

Polytechnic University
333 Jay Street
Brooklyn, New York 11201

1. INTRODUCTION

This lecture attempts to forecast into the first
decade of the 21st century new developments, expansions and
targets of polymer science and engineering. The emphasis
shall not be led on the prediction of quantities, production
facilities and prices of the large "commodity plastics" but
rather on the probable appearance of new molecules and processing techniques which are expected to open up new domains
of properties and applications.

Progress in technical sciences and engineering
emerges, in general, from three roots. (1)

1. Unexpected and surprising observations and
ideas produce a sort of <u>revolution</u> which abruptly changes
thinking and working and opens up wide new vistas and uses.

2. As soon as the "State of the Art" in a certain
field was reached a substantial level there offer themselves
"automatically" numerous opportunities for improvements and
modifications which lead to a process of slow <u>evolution</u>.
Some of such "natural" developments eventually add up to a
significant and lasting progress.

3. In the course of time new targets, markets and
demands become apparent and attractive. Some of them are of
general public importance such as safety, energy saving and
environmental protection; others come from new large scale
technical projects such as the Fusion Reaction, Outer Space
Technology, Photolysis of water into hydrogen and oxygen and
others.

Obviously there is no chance to predict any <u>revolu</u>
<u>tionary</u> events for the next decades but they may well occur.

Remember, in the past, the impact of Staudinger, Carothers, Ziegler, Natta, Pauling and Flory.

2. SYSTEMATIC EXPANSION OF THE ART

If one attempts to venture a glimpse into the future it is appropriate to take a good look at the chances for the synthesis of new molecules because they always have been the start for innovations.

There are, at present, two preparative techniques of special predominance: the Ziegler-Natta technology and the Group Transfer Polymerization. Both methods provide, essentially, the same advantages in the control of the polymer produced and, somehow complement each other, because the first is restricted to non polar monomers - olefins - whereas the other operates with polar substances particularly those which possess a carbon-oxygen or carbon nitrogen multiple bond.

The classical methods for addition polymerization use as initiators free radicals (unpaired electrons) and positive or negative ions-cationic and anionic polymerization. In all these cases the chain is started at random in a fluid (gaseous or liquid) environment and the "growing active chains and captures" additional units without any control from the ambient random monomer supply. The consequences are: irregularities of the chain structure - head to tail, head to head, branching (migration of H atoms off the chain end) and the necessity of a termination step. In Ziegler-Natta the chain is started by the addition of the monomer - e.g. ethylene - to a bimetallic complex - e.g. $TiCl_4$. This adduct represents something like a "trap" for the next monomer which is inserted between the bimetallic complex and the already attached monomer. This process continues (Ziegler's Aufbau Reaction) and the chain grows out of the complex like a "hair grows out of its root". Each inserted monomer is first "complexed" with the catalyst and is, hereby, controlled in its enchainment - stereoregulation. If there is no more monomer available the chain remains attached to the root - living polymer - and the growth may be continued with another monomer - chain segmentation. Optionally it is possible to separate the chain from its "root" in a controlled manner - choice of endgroups.

The Ziegler-Natta catalysts are now in their fourth generation with spectacular improvements in yield, control of molecular weight, stereoregularity, copolymer composition and stability against impurities.

There are coming along new types of LL co- and terpolymers, segmented copolymers of ethylene with propylene, butene, hexane octene and 4-methylpentene, a series of new electrically conducting polyconjugated species including linear poly aromatics. (2)

The other catalytic system, which is capable to revolutionize polymer synthesis provides for the same advantages - living polymer character, controlled molecular weight distribution, stereoregulation and choice of endgroups - but this time for polar monomers. It has been disclosed by a research team of the DuPont Company, together with Dr. Barry M.

Trost as Group Transfer Polymerization. (3)   Again, as in the
case of Ziegler-Natta, the chain growth is initiated by a complex - this time by a trimethylsilylketene acetal which adds
the first monomer, for instance, methyl methacrylate.   The
next monomer is then inserted into this complex so that the
initiating silyl group is continuously transfered to the
beginning of the growing chain - "which grows like a hair from
its root".

     Figure 1 indicates how such a chain is built uo from
the initiating group "i" and the monomers "m".

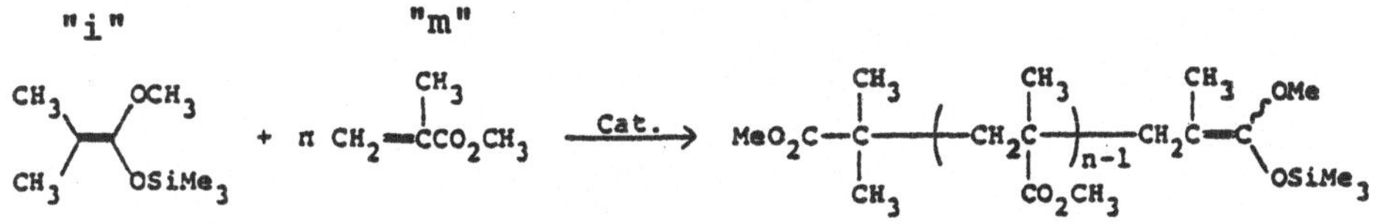

$$\underset{\text{"i"}}{\overset{CH_3}{\underset{CH_3}{>}}C=C\overset{OCH_3}{\underset{OSiMe_3}{<}}} \; + \; \underset{\text{"m"}}{n\,CH_2=C\overset{CH_3}{}CO_2CH_3} \quad \xrightarrow{\text{Cat.}} \quad MeO_2C-\underset{CH_3}{\overset{CH_3}{C}}-\left(-CH_2\underset{CO_2CH_3}{\overset{CH_3}{C}}-\right)_{n-1}-CH_2\underset{}{\overset{CH_3}{C}}-C\overset{OMe}{\underset{OSiMe_3}{<}}$$

FIGURE 1

GROUP TRANSFER POLYMERIZATION OF METHYLMETHACRYLATE

     The capacity to prepare acrylic polymers, copoly-
mers and block polymers with stereoregulation, narrow MW
distribution and controlled endgroups is obviously of considerable interest for the production of various types of
plastic windows for cars and planes and for a variety of
coatings, adhesives and lacquors.   However, it has presently
assumed even greater significance because of its connection
with the rapidly developing technique of light telecommuni-
cations. (4)

     Here the use of ultrapure polyacrylic monofils such
as Pifax - has been pioneered by the DuPont Company for short
range signal transmission and their use is now tested in
several research laboratories.

     Another vast domain for new and highly useful expan-
sions of existing polymer technology is is all branches of
transportation and building construction.   Presently this
field is dominated by metals - mainly steel and aluminum -
and ceramics - mainly concrete and glass.   All these materials have their advantages - rigidity, strength and thermal
stability but each of them has also its drawbacks.   Steel is
rigid, strong, tough and thermally stable but it is heavy
and corrosive; aluminum is much lighter but lower softening,
mechanically weaker and even more sensitive to corrosive
action of ambient conditions.   Concrete and glass are rigid
and thermally stable but they are brittle and corrosive.
Thus, all traditional large scale building materials need
modifications and improvements in certain directions.

     This had led, some 20 years ago, to the introduction
and development of the large family of specialty polymers
which today, are produced at a rate of about one million tons
a year and represent a rapidly advancing art of great fundamental interest and practical importance.

     Where are we now in this field and what logical
evolutions may be expected in the near future?

The basic design of these new building materials - the _fiber_ _reinforced_ _plastic_ _composites_ (5) - represents a combination of a rigid, strong and thermostable fiber with a thermoplastic or thermosetting matrix to give a new family of processible, durable and versatile construction materials. Intrinsically these composites consist of two essential ingredients: a _fiber_ and a _plastic_. Each of these two components is available in a large variety of individual materials; Table 1 contains a list of fibers which are used for the manufacture of such composites and Table 2 enumerates the most important plastics which are presently forming the matrices in high performance composites.

TABLE 1

Mechanical and Thermal Properties of some of the most widely used reinforcing fibers

Material	Range of tensile modulus psi	Range of tensile strength psi	Temperature of softening or decomposition °C
Carbon fiber high modulus	80×10^6	2.8×10^5	above 3000
Carbon fiber high strength	40×10^6	up to 4.1×10^5	above 3000
Al_2O_3 fiber	70×10^6	1.1×10^6	above 2300
Aramid 49 high modulus	up to 20×10^6	around 1.0×10^5	around 500
Aramid 29 high strenth	up to 8×10^6	around 1.0×10^5	around 500
Superdrawn Polyethylene	4×10^6	up to 0.7×10^5	120

TABLE 2

RESINS FOR COMPOSITES

THERMOPLASTIC
POLYOLEFINS
POLY - ESTERS, - AMIDES, - CARBONATES
-AMIDEIMIDES, - SULFONES, SULFIDES
NUMEROUS POLYBLENDS OF THEM
THERMOSETTING
EPOXIES WITH ALL KINDS OF BISPHENOLS

BISMALEIIMIDES WITH VARIOUS DIAMINES
STYRYL PYRIDINES AND COLLIDINES
BENZYLALCOHOL - PHENOLIC - COMPOSITIONS

When the first attempts were made to use composites in the construction of automobiles the designers began with the replacement of parts which were only of secondary importance for the functioning of the vehicle: bumpers, motor hoods, lid of trunk compartments, doors, etc. With the appearance of higher performance composites more integral and critical parts are also converted from metal to plastics, such as the wheels, brakes, springs, differentials and transmission shafts.

Very recently one has even started to use selected specialty plastics in the construction of the <u>engine itself</u> - an attempt which is bound to have great importance in the future.

Figure 2 shows the Polimotor Lola T-616. This engine was developed by Polimotor Research, Inc. of New Jersey. Amoto Chemicals Company is the principal sponsor. (6)

FIGURE 2

It is made predominantly of reinforced plastics, notable high performance Torlon. In its third race, the T-616 was still running strong when it finished under the checkered flag during the New York 500 endurance race at Watkins Glen.

What was a desirable convenience for the building of automobiles - rigidity, toughness and easy processibility at low specific gravity - was a necessity for the construction of airplanes. After all - they came from light weight materials - wood, canvas, bamboo and resin impregnated paper - and one was forced to use heavier materials: steel, aluminum and eventually titanium as the vehicles became larger, higher flying and faster. Evidently, it would be nice if one could revert this trend by the use of the new high performance composites. In fact, already during World War One, all then available synthetics - cellulose derivatives, phenolics and amio resins - were used in fighter and reconnaissance planes as much as possible. During the following 50 years this trend persisted and was intensified through the necessity to design and build transatmospheric vehicles and even space craft for orbital and transorbital missions. Figure 3 shows how many parts of the Boeing 757 are now replaced by composites and Figure 4 presents the same infor- mation for the Boeing 767 where even more light weight composites are used. Still more replacement is carried on in the European Airbus 300 and 310.

Obviously this is just a beginning but has already reduced the weight of a wide body plane by several tons. Each ton represents an attractive option: either to put on the plane more passengers or the equivalent amount of freight or to save fuel the cost of which accounted in 1980 for 55 percent of the direct cost of operating an airline.

Altogether the existing art is very mobile and invites at many instances pioneering activities which shall gradually move forward the boundaries of our knowledge and know-how.

3.   Demands which stimulate the invention of products and processes

Demands, needs and trends of many kinds have always stimulated progress in science and technology. There are general public demands such as the cry for a "cleaner and safer world" which brought about great improvements in the protection of the environment and in safety measures against fire, intoxixation and contagious diseases. The opening up of new attractive markets has always been an important incentive for the development of new materials and processes. Finally, new vistas and projects of major proportions like the photolysis of water, the fusion reaction or the conquest of space are generating initiatives for larger, sustained efforts leading to a substantially higher level of understanding and know-how.

Let us take a look at what is likely to happen along these lines during the approaching end of this century. The contamination of our atmosphere is caused by many sources. In the USA one of the worst are the effluents products by automobiles - many of them - as they are operated with increasing numbers in all industrial countries. Much is already being done to ameliorate these unsafe and sometimes scandalous conditions but a considerable change for the better could only be expected if hydrogen could be used as fuel or if the vehicles could be driven electrically. Much work is done in both directions but it has little to do with the use of polymers and is, therefore, somewhat outside of the scope of this lecture.

But, besides this ubiquitous contaminating action through moving vehicles there exist several large scale industrial activities which represent local centers of particularly bad influence on the ambient environment. In all cases strong efforts are made to minimize this influence and, frequently, polymers - already known or yet to be developed - are of great help.

In an oil refinery, for instance, heat is needed in enormous quantities to crack the crude and to separate the resulting chemicals - lub-, fuel - and diesel oil together with the lighter gasoline fractions. It is also needed to produce electricity for pumping, compressing, evacuating and for the entire internal operation of the refinery. Heat - in large amounts and at high temperatures is always dangerous for the environment, particularly if it is produced by coal or oil - as is often the case. Search for catalysts which allow to crack, isomerize, disproportionate and, eventually, polymerize hydrocarbons at lower temperatures are a "hot" research tar-

FIGURE 3

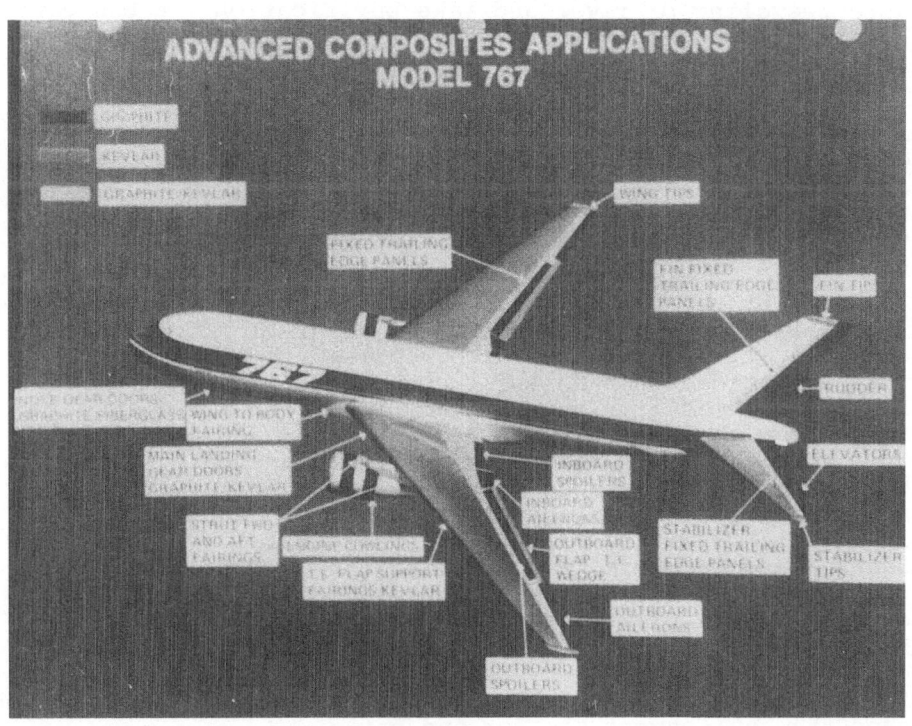

FIGURE 4

get. Beyond that a system of new separation techniques for gases - H_2, CO_2, CH_4 and for liquids - benzene, toluene, hexane - is now being developed and introduced; it is based on the different rate of diffusion of these materials because of the size, shape and polarity of their molecules. (7)

The essential feature for these operations is the use of a membrane in the form of a film or a hollow fiber consisting of a very thin and dense separation layer and a thick strong and porous support. The polymer which forms the two layers may be the same; then the difference in their structure and properties depends on the way of layer formation. For most cases, however, the separation layer and the support are made of different polymeric materials. Both of them must be tough and abrasion resistant and must not be attacked by any material with which they may come in contact -liquid or gaseous; they also must be durable enough to guarantee a long (several years) service life. For years the separation layer has been a cellulose derivative - acetate, buterate or copolymer; recently other thermoplastics are being used: aramides, arylates, polyphenylene sulfides or sulfones and others. In the future any of the new "internal" vinyls and acrylics will be tested for use. This layer should be very thin - preferably less than one micron and absolutely free of pinholes. The support is much more open, sometimes even spongy and is as thick as the application conditions permit - up to 2 or 3 mils (50 to 75 microns).

Presently there are already large separation units in satisfactory operation; million gallons of water, gas and solvent are purified per day. In the future most water recovery processes like desalination, urban and industrial waste recycling or river and lake purification are bound to be carried out with membranes for ultrafiltration and reverse osmosis.

At the same time air pollution is going to be minimized by leading the hot gases which contain ultrafine aggressive particles and acidic or basic components through fine filters made of thermostable fabrics. Much finer and stronger bags made of aramides, carbon fibers and polyimides will be available to arrive at a faster and superior gas purification. (8)

All this is now being used and will be used more frequently and on a higher level of sophistication in the future in all large scale industries; where high temperatures are needed such as steel, concrete, glass and paper. In the paper industry there is now initiated a radical change from the presently used aggressive chemicals Mg, Ca and S to much milder techniques for the separation of cellulose from lignin using only organic ingredients like alcohols and esters. (9) No sulfurous compounds can escape into the air and all resulting solid products - cellulose, lignin and sugars - are recovered with no toxic or obnoxious effluents. This method is still in its infancy but pilot plants are now being operated at various locations and by the year 2000 there will certainly be many full size mills using this new technology because it arrives at the same or better quality of cellulose, lignin and hemicelluloses and, at the same time keeps the environment cleaner and safer.

Recycling of waste is another effort which is finding more and more serious attention in agriculture, forestry, industry and urban life. Everywhere there exists the tendency to recover at least the caloric equivalent of the material by combustion under appropriate conditions avoiding the formation of "fly-ash" and the release of toxic gases into the atmosphere. Thermostable polymeric, filter bags and non woven mats are now used with good results and will be applied on a much larger scale in the future.

On top of these relatively modest demands for environmental protection, energy saving, recycling and the filling of occasional market niches there exist several much larger problems which somehow must be solved, at least to a certain extent.

It is probable that by the middle of the next century the Natural Resources for the production of energy - coal, oil and gas - will become so scarce that they have to be exclusively used for the provision of organic chemical raw materials to produce synthetic fibers, rubbers and plastics. Also the renewable material sources from forestry, agriculture and aquiculture will be strictly excluded from being used as energy sources through combustion. They will be at that point and will remain, for all time, our only source or organic matter.

Presently, the combustion of these materials provides for more than 80% of our energy needs - hydroelectric and atomic power through the fission reaction - being the only alternative energy sources which count.

The principles of the Fusion Reaction are well known and its uncontrolled and explosive performance has culminated in the Hydrogen Bomb. Elaborate research has been going on during the last 20 years in numerous laboratories to control this type of nuclear reaction and to arrive at the delivery of thermal energy on a level which will permit its further practical use. Not much progress has been made and it almost appears that we still have to look for a new fundamental principle to open for us the way to a large scale practical use of "synthetic" nuclear power.

The same thing seems to be true for another hopeful process for power production through the provision of hydrogen - through the photolysis of water. This process goes on every day under our eyes in all plants; they are capable to split water into its elements and to react the hydrogen with the CO_2 to produce formaldehyde CH_2O and from it the entire world of organic chemical compounds. This reaction takes place under ambient conditions with a high yield of individual chemical steps pro absorbed photon from the sunlight. Intensive research, going on now for several decades has revealed that the process involves a complex system of enzymes for the replacement of which we need to know some new fundamental principle of bio- and photoengineering.

Thus it is difficult to predict whether and at what time these two "macropower" sources will be available for large scale practical use. What remains to us as a realistically foreseeable method to produce energy (electricity) directly from sunlight is the use of solar cells at the surface of the earth or, much better, in space at an order of magnitude of several hundred gigawatt. The direct conversion of sunlight into electricity is the energy source for all satellites and has been - on a small scale - very successful. Estimates have been made that by 2020 the energy demand of the then existing population will be about 2000 gigawatt. Its satisfaction from terrestrial resources will not be any more possible because they will not exist at that time. Hence a number of large Solar Power Stations are now in a planning and designing state. (10)

REFERENCES

(1)   This report is an extension of a lecture given at the K-86 Pre Show Conference in Duesseldorf on November 4, 1986.

(2)   Compare here particularly: J.C.W. Chien "Polyacetylene" Acad. Press, New York 1984; also the "Marvel-Monsanto Lectures" 1986, at the University of Arizona presented by Professor Gerhard Wegner and "The Polaroid Corporation" data sheet on TCP-117 film and coating 1986.

(3)   B.C. Anderson et al. Macromolecules 14, 1599 (1981) C.D. Andrews and A. Vatvars: ibidem 14, 1603 (1981) O.W. Webster et al. ibidem "In press."

(4)   "Optical Fiber Telecommunication" S.E. Miller and A.G. Cynoweth, Acad. Press, New York 1979; particularly the last 6 pages.

(5)   The Literature on high performance composites is very large; reference is given here only to two recent books: "Polymer Composites," editor: B. Sedlacek; de Ganyler; New York 1986 and Erich Fitzer, "Carbon Fibers and their Composites" Springer-Verlag - New York 1985.

(6)   Technical Information Sheet on the Polimotor Torlon 8 Engine of Amoco Chemicals Company, 1986; 200 E. Randolph Drive, Chicago Illinois 60601.

(7)   See, for instance, the book on "Membranes" O.J. Sweeting, Yale University, "The Science and Technology of Polymer Films" Interscience, N.Y. 1968, also, R. Schmid in H. Batzer "Polymere Werkstoff" page 211 et seqn. Thieme Verlag, New York 1985; also elaborate literature on "Permasep" of the DuPont Company "Permasep" Products, Glasgow Site; Wilmington, Delaware and the present State of the Art and many attempts for further improvements can also be found in the current issues of "Packaging Engineering."

(8)   Compare the elaborate literature on Waste Incineration Plants of the Signal Division of Allied Chemical Corp Morristown, New Jersey.

(9)   See, for instance: Fogaraty el al. USO 3 887 426 - June 19, 1975 Diehold et al. USP 4 100 016 - July 19, 1978

      Also: R. Katzen et al. Chem Eng Progress 21, 62 (1980) and Austrian Patent Appl. A-3483/ 85, filed on November 29, 1985.

(10) See the book of Farrington Daniels on "Direct Use of the Sun's Energy; Yale University Press, New Haven 1974. Also H.F. Hildebrand and L.L. Van Hall "Power from Heliostate: Science, Sept. 16, 1977 and particularly, P.E. Glaser, "Power Satellites Development" J. Astr. Science, April 1978.

INDEX